THE PINEWOOD NEMATODE, *BURSAPHELENCHUS XYLOPHILUS*

PROCEEDINGS OF AN INTERNATIONAL WORKSHOP, UNIVERSITY OF ÉVORA, PORTUGAL, AUGUST 20–22, 2001

THE PINEWOOD NEMATODE, *BURSAPHELENCHUS XYLOPHILUS*

PROCEEDINGS OF AN INTERNATIONAL WORKSHOP,
UNIVERSITY OF ÉVORA, PORTUGAL,
AUGUST 20–22, 2001

M. Mota and P. Vieira (Editors)

NEMATOLOGY MONOGRAPHS AND PERSPECTIVES

VOLUME 1

BRILL
LEIDEN – BOSTON
2004

This book is printed on acid-free paper.

Library of Congress Cataloging-in-Publication Data

The Library of Congress Cataloging-in-Publication Data is available from the Publisher.

Cover illustration: *Bursaphelenchus xylophilus*, male tail. Part of Fig. 3 on page 4. Photo by A. Catarina Penas.

ISBN 90 04 13267 8

PRINTED IN THE NETHERLANDS

Sponsors:

UNIVERSIDADE
DE EVORA

DELTA
Café

região de turismo
ÉVORA

FUNDAÇÃO
ORIENTE

INSTITUTO
DA VINHA
E DO VINHO

FCT
Fundação para a Ciência e a Tecnologia
MINISTÉRIO DA CIÊNCIA E DA TECNOLOGIA

BANCO
ESPIRITO
SANTO

Miolo de Pinhão
Pinheiro
Francisco P. Cecílio & Filhos, Lda.
PORTUGAL

FUNDAÇÃO
LUSO-AMERICANA

Vinhos do Alentejo

évora
câmara municipal

CONTENTS

Morphology and molecular methods of identification: Session organisers H. Braasch, W. Burgermeister

Ecology and epidemology; quarantine issues: Session organisers D. Bergdahl, K. Futai

Physiology, resistance and histopathology: Session organisers Yu-Yang, K. Kuroda

Biology of PWN and relationships to its cerambycid vectors: Session organisers M. Linit, B.J. Yang

Control methods: Session organisers B.J. Yang, K. Nakamura

Introduction

This is an extraordinary example of an unusually rapid and effective collaboration between science and public policy.

New risks are commonly described as disruptive of the normal ways in which social response is organised. However, and under special circumstances, risk may trigger a higher level of collaboration and a more effective involvement of scientists, technicians, politicians and administrators. Risk can be perceived as a challenge and as an opportunity.

This is the story I am proud of having modestly been able to contribute to. Prof. Manuel Mota and his colleagues have taken the lead and the scientific responsibility for both announcing the bad news and helping the main national organisations to solve the problem. They have contributed to the building up of a renewed networking of university and national laboratories as well as of administrators and inspection bodies.

The continuity of this new network is a critical question for the future. Responding to a new risk has provided the opportunity and the common drive for a major collaborative effort. Science has been able to help triggering this very effective collective action and to gain a new legitimacy. We hope that as a result of this process science has also become more embedded in social life and is now increasingly respected.

And we must praise all those who have played a significant role in the discovery of the (unfortunate) emergence of the pinewood nematode in Portugal and in the subsequent fight against its spreading for their tenacity.

José Mariano Gago
Minister of Science and Technology of Portugal (1995-2002)

Nematology Monographs & Perspectives, 2003, Vol. 1, xiii-xiv

Introduction

Pine forests represent an extraordinary natural resource worldwide, with a wide variety of products such as paper, lumber, wood chips, resin, *etc.* For many countries, pine wood represents a major portion of their gross national product. Commerce of wood products between countries and between continents plays a vital role in the world economy, with direct and indirect impacts on jobs, construction and other human activities.

Among the major pests and pathogens of conifers, the pine wood nematode, *Bursaphelenchus xylophilus,* constitutes a serious concern in many parts of the world, but especially in Asia where it has ravaged many pine stands, for example, in Japan. The nematode is apparently native to North America, where it has co-evolved with the local conifer species, thus causing little damage to the stands and to the national economies of Canada and the United States.

Recently, in 1999, this pathogen, classified by the European Plant Protection Organization (EPPO) as an A1 organism (very high risk), was detected in Portugal and in Europe for the first time. Following detection, the members of the research team involved immediately informed the national authorities, providing the scientific evidence of its presence. The national authorities promptly communicated this fact to the European Union (EU) agency in charge of plant quarantine issues.

A major effort was developed among researchers, technicians and political personnel in Portugal, in order to study the nematode and its vector (*Monochamus galloprovincialis*), and to implement control measures. The nematode has so far been contained within the affected area of the Setúbal Peninsula, 30 km south of Lisbon. The Portuguese authorities and the EU officials have worked jointly to avoid the spread of the pathogen. Researchers of several institutions such as the Estação Agronómica Nacional, and the Estação Florestal Nacional (belonging to the Ministry of Agriculture), and the University of Évora, have proceeded with basic studies on the biology and ecology of both the nematode and the insect.

In 2001, it was timely to meet with the international community working on this subject and exchange views on recent research as well as control measures. The NemaLab/ICAM, University of Évora, decided to promote the organisation of an international workshop that took place in Évora between August 20 and 22, attended by nearly 50 researchers from 14 countries worldwide. The large majority of the international experts were present, and their presentations are published in this volume.

Évora, August 2003

Manuel M. Mota & Paulo Vieira
NemaLab/ICAM, Department of Biology
University of Évora

Nematology Monographs & Perspectives, 2003, Vol. 1, 1-5

Discovery of pine wood nematode in Portugal and in Europe

Manuel M. MOTA [1], Luis BONIFÁCIO [3], Ma Antónia BRAVO [2], Pedro NAVES [3], Ana Catarina PENAS [2], Joana PIRES [2], Edmundo SOUSA [3] and Paulo VIEIRA [1]

[1] *NemaLab/ICAM, Departamento de Biologia, Universidade de Évora, 7000 Évora, Portugal*
[2] *Departamento de Protecção de Plantas, Estação Agronómica Nacional/INIA, Qta. do Marquês, 2780 Oeiras, Portugal*
[3] *Departamento de Protecção Florestal, Estação Florestal Nacional/INIA, Qta. do Marquês, 2780 Oeiras, Portugal*

Summary – Pine wood nematode, *Bursaphelenchus xylophilus*, was first reported for Portugal (and Europe) in 1999. The importance of this discovery and economic impact are discussed. Details of the ongoing research of the nematode and its vector, *Monochamus galloprovincialis*, are provided, and are mainly on the morphology and molecular biology of the nematode and the bioecology of the vector and its relationship with *B. xylophilus*. Pine products play an important role in the Portuguese economy. The total area of forest trees in Portugal is approximately 3×10^6 ha, of which *Pinus* species occupy roughly 1.25×10^6 ha. Pine products include lumber, resin, pulp and pine seed, all of which are very important economic products in our country. The central region of Portugal contains the largest continuous area of maritime pine, *Pinus pinaster*, in Europe. On several occasions, Portuguese researchers have pointed to the dangers of the possible presence of the pine wood nematode (PWN), *Bursaphelenchus xylophilus*, an A1 quarantine pest, according to EPPO, in Portugal (Macara, 1994).

Detection of PWN in Portugal

The presence of *B. xylophilus*, the pine wood nematode, in Portugal was first reported in May 1999 (Mota *et al.*, 1999) and is the first record of this extremely damaging organism within the European Union. Other

Fig. 1. *Declining pines (*Pinus pinaster*) near Lisbon, Portugal with symptoms of pine wood nematode infestation.*

Bursaphelenchus species have been reported in Europe, associated with wilted pines. For a more detailed description on this subject, see Mota (2002). The discovery was a result of a joint research effort between the University of Évora and the research organisations of the Ministry of Agriculture of Portugal (INIA/EAN and EFN). Following detection, the Portuguese authorities promptly informed the European Union (EU) of the presence of pine wood nematode, in mid-1999. A task force (GANP) was immediately established, followed by a national programme of survey and control of the pine wood nematode (PROLUNP) (Serrão, 2001). From 1999-2001, several EU phytosanitary inspection teams have been able to verify the results of the PROLUNP actions. The nematode has been shown to be contained in the Setúbal peninsula, approximately 30 km south east of Lisbon, where symptoms may be occasionally observed (Fig. 1). At present, *P. pinaster* is the only affected species in mixed stands with *P. pinea* and *P. halepensis*.

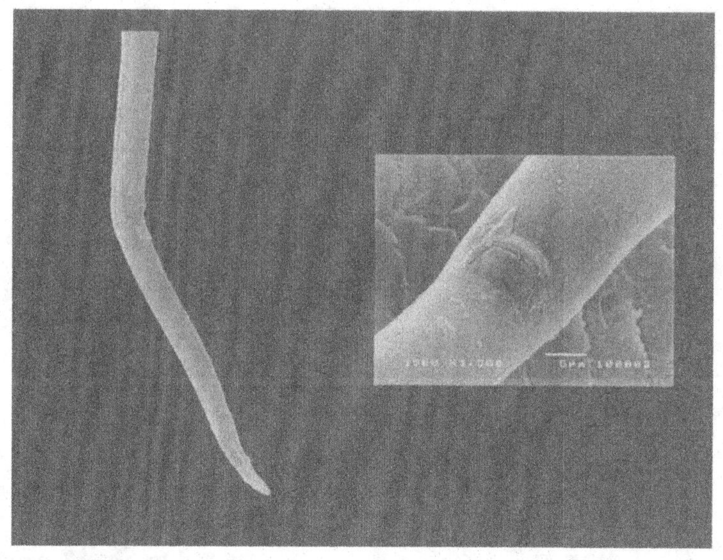

Fig. 2. *Scanning electron micrographs of* Bursaphelenchus xylophilus *female tail and vulval flap (inset).*

Research

Ongoing research has included morphological and biometrical observations of populations of *B. xylophilus* from the Setúbal peninsula, including scanning electron microscopy (SEM) observations of the female tail and vulval flap (Fig. 2) and of male tail and spicules (Fig. 3). Male spicules are of particular interest when observed by SEM, providing greater detail of its typical morphology and diagnostic features. Other *Bursaphelenchus* species occurring in pine wood, as well as the extraction methods involved, have also been studied (Mota, 2002; Penas *et al.*, 2002a, b). Recently, a CD-rom containing relevant information on PWN has been released (Vieira *et al.*, 2001) providing researchers with the original description of all species within the genus *Bursaphelenchus*.

Of major interest is the molecular characterisation of the nematode DNA and, in particular, the ITS region of rDNA (Hoyer *et al.*, 1998; Mota *et al.*, 1999). Ongoing research is focusing on characterisation of rDNA from different populations of *B. xylophilus*, as well as of different *Bursaphelenchus* species occurring in Portugal (Penas *et al.*, 2002b).

Although the nematode has occasionally been reported in the literature on other beetle genera, *Monochamus* seems to be the only effective

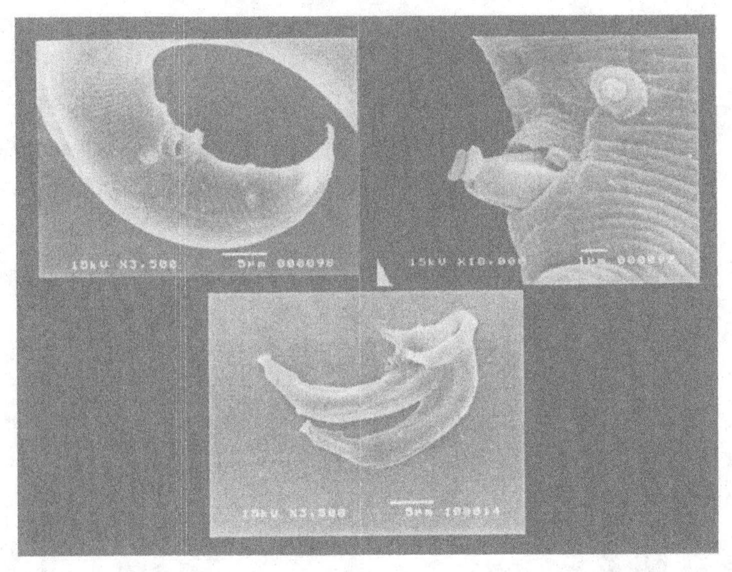

Fig. 3. *Scanning electron micrographs of* Bursaphelenchus xylophilus *male tail and spicules.*

vector for spread of PWN. These longhorn beetles also transmit *B. mucronatus,* which is distributed in Europe and very similar to *B. xylophilus*, but less pathogenic. Recently, *M. galloprovincialis* has been proven to be the vector of *B. xylophilus* in Portugal (Sousa *et al.*, 2001). Ongoing research regarding this vector focuses mainly on its bioecology and control.

Nothing is known so far about the origin of the *B. xylophilus* strain detected in Portugal. However, increased commercial movement in the last few years from Macau (China) to Portugal, in the form of ship containers with large numbers of wooden crates, establishes an interesting testable hypothesis. Other possibilities include wooden crates from North America, as containers of automobile parts. A new research project, involving eight partners from the EU, will soon integrate the efforts of survey and studies from six EU countries.

Acknowledgements

The authors thank Dr Ana Luisa Diogo for her SEM image of an extracted pair of *B. xylophilus* spicules (Fig. 3). This work has been funded by FCT (Fundação Ciencia e Tecnologia), under project Praxis XXI (11 189/98) and POCTI (32 619/99).

References

EVANS, H.F., MCNAMARA, D.G., BRAASCH, H., CHADOEUF, J. & MAGNUSSON, C. (1996). Pest Risk Analysis (PRA) for the territories of the European Union (as PRA area) on *Bursaphelenchus xylophilus* and its vectors in the genus *Monochamus*. *EPPO Bulletin* 26, 199-249.

HOYER, U. VON, BURGERMEISTER, W. & BRAASCH, H. (1998). Identification of *Bursaphelenchus* species (Nematoda, Aphelenchoididae) on the basis of amplified ribosomal DNA (ITS-RFLP). *Nachrichtenblatt des Deutschen Pflanzenschutzdienstes* 50, 273-277.

MACARA, A.M. (1994). [Nematofauna associada a plantas florestais em Portugal (1987-1992).] *Revista de Ciências Agrárias* 17, 77-126.

MOTA, M. (2002). Occurrence of the pinewood nematode, *Bursaphelenchus xylophilus* in Portugal and perspectives of the disease spread in Europe. *Nematology* 4, 124. [Abstr.]

MOTA, M.M., BRAASCH, H., BRAVO, M.A., PENAS, A.C., BURGERMEISTER, W., METGE, K. & SOUSA, H. (1999). First report of *Bursaphelenchus xylophilus* in Portugal and in Europe. *Nematology* 1, 727-734.

PENAS, A.C., DIAS, L.S. & MOTA, M. (2002). Precision and selection of extraction methods of aphelenchid nematodes from maritime pine wood, *Pinus pinaster* L. *Journal of Nematology* 34, 62-65.

PENAS, A.C., BRAVO, M.A., PIRES, J. & MOTA, M. *Bursaphelenchus* species found in maritime pine in Portugal. *Nematology* 4, 273. [Abstr.]

SERRÃO, M. (2003). Eradication programme for the pine wood nematode in Portugal. In: Mota, M. & Vieira, P. (Eds). *Proceedings of the pine wood nematode symposium, Evora, Portugal. Nematology Monographs and Perspectives* 1, 7. [Abstr.]

SOUSA, E., BRAVO, M.A., PIRES, J., NAVES, P., PENAS, A.C., BONIFÁCIO, L. & MOTA, M.M. (2000). *Bursaphelenchus xylophilus* (Nematoda; Aphelenchoididae) associated with *Monochamus galloprovincialis* (Coleoptera; Cerambycidae) in Portugal. *Nematology* 3, 89-91.

VIEIRA, P., EISENBACK, J.D. & MOTA, M.M. (2003). PWN-CD: a taxonomic database for the pinewood nematode *Bursaphelenchus xylophilus*, and other *Bursaphelenchus* species. In: Mota, M. & Vieira, P. (Eds). *Proceedings of the pine wood nematode symposium, Evora, Portugal. Nematology Monographs and Perspectives* 1, 165. [Abstr.]

Eradication programme for the pine wood nematode in Portugal

Miguel SERRÃO

Secretaria de Estado do Desenvolvimento Rural, Ministério da Agricultura, Desenvolvimento Rural e Pescas, Terreiro do Paço, Lisboa, Portugal

Summary – The pine wood nematode (PWN) was first detected in Portugal in May 1999. Following detection, Portuguese authorities initiated the implementation of eradication measures, aware of the seriousness of the situation and the urgent need to act. The phytosanitary risk and possible related economic and commercial consequences led to the involvement of all sectors affected by the problem to try to control and eradicate PWN. A task force from the follow-up group (GANP) established by the Secretary of State for Rural Development established a national eradication programme (PROLUNP) to: *i*) contain PWN within the initial geographic limits; *ii*) implement eradication measures; and *iii*) monitor PWN at a national level. PROLUNP integrates different sub programmes and actions, involving different official services together with the private sector, and is co-ordinated by the Secretary of State for Rural Development. Strategies and organisation as well as actions and methodologies are presented along with the results. The main constraints and solutions encountered during the process are discussed. The result of the PROLUNP actions foresees a scenario where the eradication of PWN from Europe may be possible in a reasonable time frame; nevertheless, an extraordinary effort has to be made to ensure the necessary resources.

Nematology Monographs & Perspectives, 2003, Vol. 1, 9-20

Pine wilt disease in Japan

Yasuharu MAMIYA

5-6-8 Kitanodai, Hachioji, Tokyo, 192-0913 Japan

Summary – Pine wilt disease caused by the pine wood nematode, *Bursaphelenchus xylophilus*, is widespread throughout Japan except for the two northernmost prefectures. The disease has caused the death of millions of pine trees (*Pinus densiflora*, *P. thunbergii* and *P. luchuensis*) every year, and an annual loss in timber of as much as 1 million m^3. The enormous increase in timber loss in the 1970s resulted in a loss of 2.4 million m^3 in 1979. This increased loss might be due to a number of reasons. The nation's economic and social development has led to drastic changes in the demand for pine timber. Consequently, attempts to control the disease were neglected, and dead trees left in the forests, resulting in an increase in the sources of infection and of vector, *Monochamus alternatus*, populations. Since 1975 the disease has spread to northern Japan. Outlying infection, resulting from the introduction of nematode-infected pine logs, is a common cause of disease occurrence in disease free areas. In warm areas such as Kyushu, where the disease occurred first in Japan in the early 20th century, the disease has destroyed many pine stands and, in most cases, they have been replaced with natural forests of broadleaved trees. In heavily damaged areas, conversion of pine forests to other tree species will be encouraged in order to keep the forest resources at a desirable level and to maintain the function of forests in land conservation. Efforts should continue to control the disease in pine forests at a primary stage of infestation in the northern regions, and other specified pine forests which play an important role for the public benefit in protecting coastline, preventing natural disaster and sustaining the aesthetic value.

The pine wilt disease induced by the pine wood nematode, *Bursaphelenchus xylophilus*, is a great threat to pine forests in Japan. Its epidemic spread and damage is so rapid and serious that pine forests over a large area have been destroyed by the disease so far (Mamiya, 1988). Since the first occurrence of pine wilt disease in Nagasaki, Kyushu, in 1905 (Yano, 1913), this devastating epidemic has spread throughout Japan except for the two northernmost prefectures, Aomori and Hokkaido, out of 47 prefectures. It has caused serious damage to the most common native pines, *Pinus densiflora*, *P. thunbergii* and *P. luchuensis*. In 2000, infested areas

Table 1. *Pine forests in Japan and damage and losses from pine wood nematode.*

	Area (10^6 ha)	Damaged pine forests		Timber loss[2] (10^6 m^3/y)	Stock volume (10^6 m^3)	Wood products (10^6 m^3)
		(10^6 ha)	% area[1]			
Total forest land	25.1	–	–	–	3500	49
Pine forest						
1979	2.6	0.65	25	2.43	312	4.2
1990	2.2	–	–	0.87	336	2.7
1999	2.1	0.58	28	0.72	356	1.4

[1] Percent of damaged areas in total pine forest.
[2] Annual loss caused by the pine wood nematode.

were estimated to be 580 000 ha, 28% of Japan's total of 2.1 million ha of pine forest (Table 1).

In 1971, the pine wood nematode was first described as the causal agent of pine mortality (Kiyohara & Tokushige, 1971). Pine sawyer, *Monochamus alternatus*, was immediately designated as a principal vector of the causal agent (Mamiya & Enda, 1972; Morimoto & Iwasaki, 1972). It took more than 60 years for the primary pathogen of the disease to be recognised. During this period efforts to control the disease were made based on entomological strategies.

History of pine wilt disease in Japan

The increase in loss of pines and extensive spread of the disease can be followed using the statistical data accumulated by the Forestry Agency of the Japanese Government (Figs 1, 2, 3).

In 1905, the first occurrence of pine wilt disease was recognised in Nagasaki. This outbreak of pine wilt disease was controlled by eradication of dead trees, which prevented the recurrence and spread of the disease in the area. In 1925, an outbreak of the disease was reported in Sasebo, 50 km north of Nagasaki. After several years, the disease caused a considerable amount of dead pine trees in the area, because no effort was made to control the disease as the infested area included land under military use. Thus an infection centre of the disease was

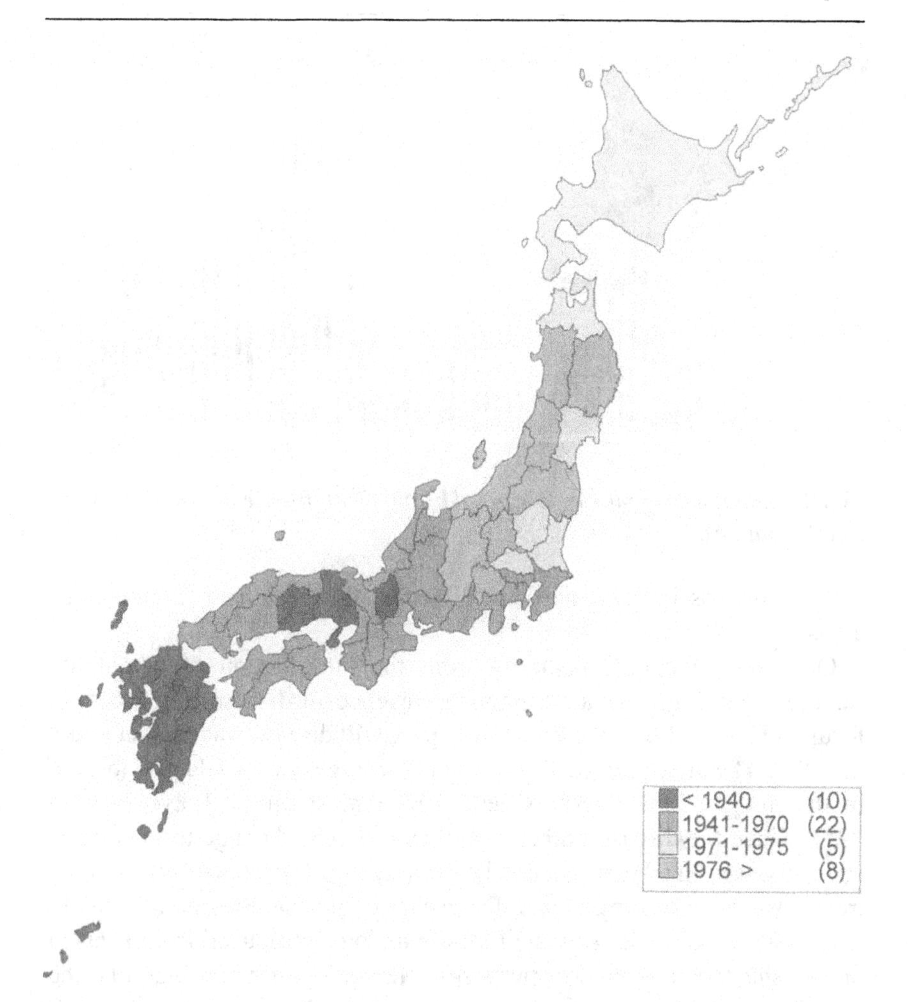

Fig. 1. *The spread of pine wilt disease in districts of Japan after the first occurrence in the 1900s.*

first established in Kyushu and the disease spread into surrounding areas during the 1930s. In 1921, pine wilt disease was found in Honshu at a locality in the Hyogo Prefecture. This was the first record of the disease outside of Kyushu.

During the 1930s, pine wilt disease extended to ten prefectures, seven of which were in Kyushu. By the 1960s, the disease areas had extended as far as Kanto in central Honshu, and included more than 30 prefectures. After 1971, the infested areas spread rapidly and extensively. Since 1975

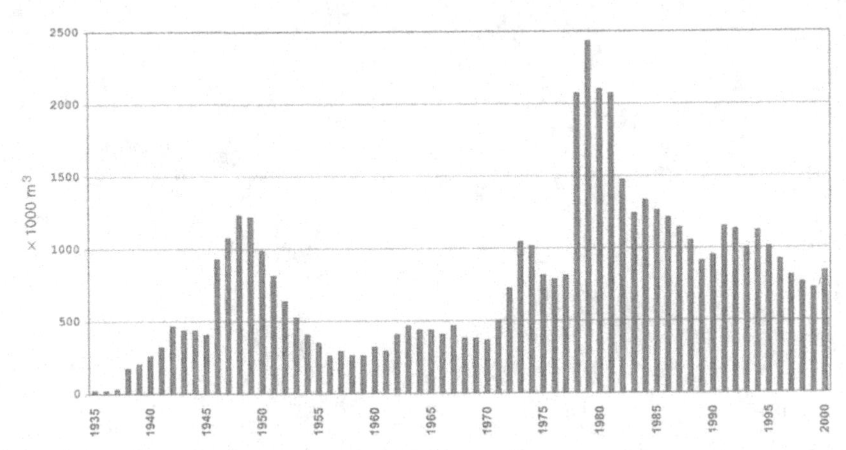

Fig. 2. *Annual loss of pine trees caused by pine wilt disease in Japan (Forestry Agency, Japan).*

the disease has invaded northern Honshu and the inland mountainous areas.

Outlying infections, resulting from the introduction of nematode-infected pine logs, are a common occurrence in disease-free areas. In Ryukyu Islands, Okinawa Prefecture, pine wilt disease was first recorded in 1973. The first record of the disease in Ogasawara Islands, located in the midst of the Pacific Ocean 1000 km south of Tokyo, was in 1978. Since the disease outbreak in these islands, damage to pine trees, *P. luchuensis*, has been absolutely devastating, and almost all the pine trees have been destroyed and disappeared from Ogasawara islands. In both cases the disease originated from pine logs introduced from Kyushu or Honshu for construction purposes. Nematode-infected logs may be introduced together with healthy pine logs from infected areas.

During the 1930s and the early 1940s, the annual loss of timber increased to 400 000 m^3. In 1948, a loss of 1.2 million m^3 resulted from abandoning attempts to control the disease during World War II. Subsequent control efforts, mostly involving eradication of dead trees, reduced the annual loss to 500 000 m^3. Timber loss was kept more or less at this level during the 1950s and 1960s. However, annual losses in the 1970s again exceeded 1 million m^3. In 1979 the heaviest annual loss of timber, 2.4 million m^3, was recorded. The increased loss might be due to one of several of the following reasons. The nation's economic and social development led to drastic changes in the demand for pine timber. The utilisation of pine timbers drastically declined. Fuel innovation

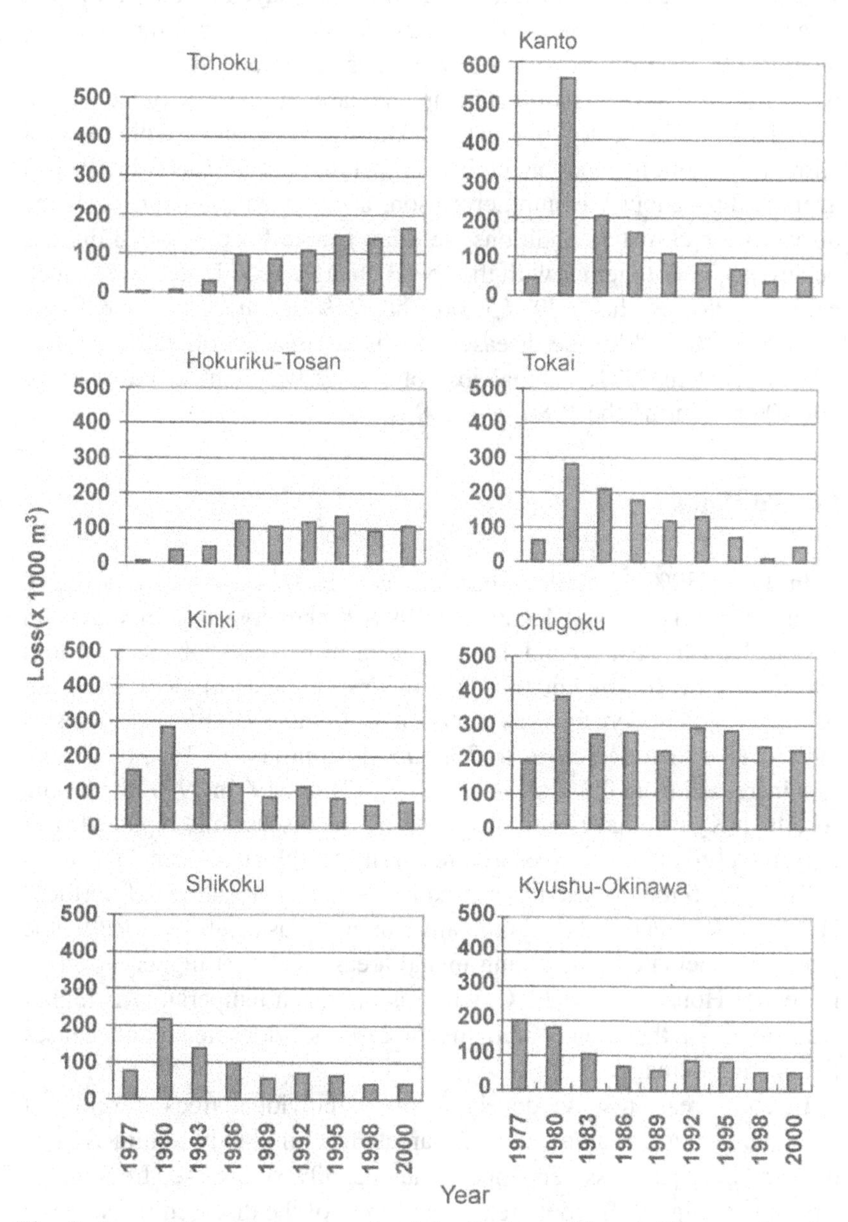

Fig. 3. *Annual loss of pine trees in eight districts of Japan (Forestry Agency, Japan).*

where wood fuels were replaced by fossil fuels played an important part in accelerating the decrease in demand for pine timber. Consequently, attempts to control the disease were neglected and dead pine trees were left in forests, resulting in an increase in sources of nematode infection and in vector population. During 1978 and 1979, unusual climatic conditions, such as higher temperatures with less rainfall than average throughout the summer season, accelerated an outbreak of the disease. The climatic conditions were favourable for promoting disease occurrence and, together with the social changes mentioned, led to such enormous losses. In 1979, 650 000 ha, 25% of the total pine forest area, was affected by the disease. This is an area 6.5 times larger than that affected in 1971. Annual loss of timber was kept at the level of 800 000 m^3 during the 1980s and 1990s.

Present status

In 1977, 80% of total timber loss was in western Japan, Kyushu, Chugoku, Shikoku, and Kinki. By 1999, timber loss had increased in eastern Honshu, Kanto and Tohoku. As a result, 40% of the total loss was in eastern Honshu and 60% was in western Japan in 1999. Decrease of timber loss in Kyushu was remarkable. It was 27% of total losses in 1977, and it then decreased to 7%. On the contrary, in Tohoku, timber loss increased from 0.2% of total loss in 1977 to 21% in 1999. Most pine forests in Kyushu have been destroyed by the disease and replaced with evergreen broad-leaved trees as a result of natural succession.

Since 1975 the disease has spread into scattered localities of northern Honshu. The mean annual temperature in the areas newly invaded by the pine wood nematode, including inland areas located at higher elevation in central Honshu, is 10-12°C, whereas the mean temperature is higher than 14°C in the area where the disease is widespread and causes devastating damage.

In cool areas, disease development in individual trees is different from those in warm areas. In warm areas, from mid-July to mid-August diseased trees progressively appear and over 90% of diseased trees die by December (Fig. 4). In cool areas, nearly half of the diseased trees do not die within the year of infection (Fig. 5). These trees die in early spring to early summer in the following year. This delayed disease development is termed biennial disease development.

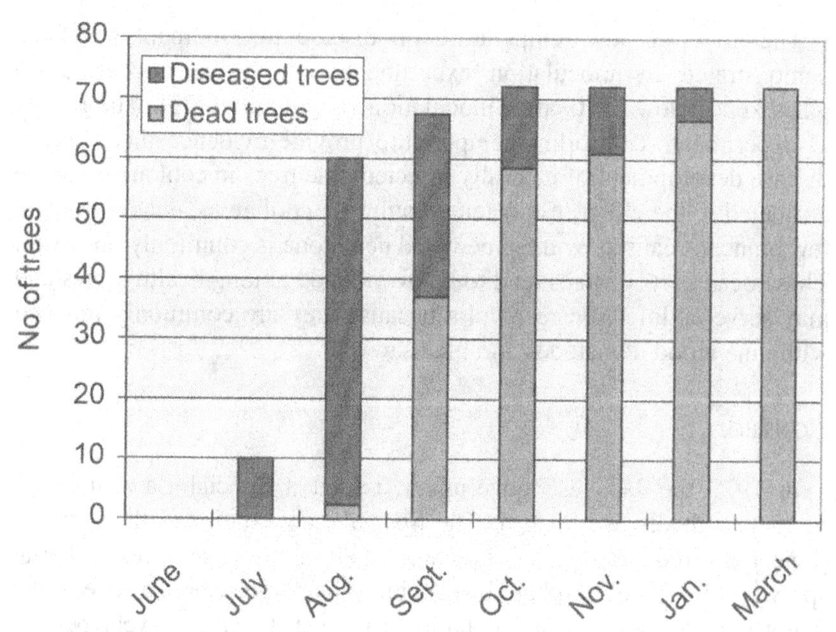

Fig. 4. *Natural infection of pine trees with the pine wood nematode and disease development. A total of 332 of 25-year-old* P. densiflora *were examined at a forest in Chiba Pref. (After Mamiya* et al., *1973.)*

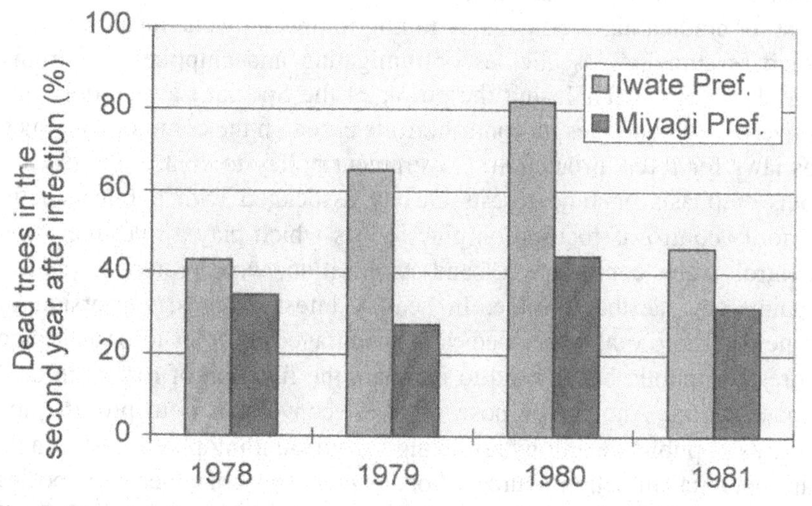

Fig. 5. *Biennial disease development of pine trees infected with the pine wood nematode in cool areas, Tohoku.*

The effect of low temperature on disease development has been demonstrated by inoculation experiments. No disease occurred on pines kept below 20°C after inoculation (Kiyohara, 1973). The results of experiments controlling temperature provide evidence that delayed disease development of naturally infected pine trees in cool areas can be attributed to the effect of low temperature. In cool areas, death of only a few branches caused by the pine wood nematode is commonly observed. These dead parts of trees tend to be overlooked among healthy trees and may serve as infection reservoirs because they are commonly infested with pine wood nematodes and pine sawyers.

CONTROL

In 1977 the Japanese government issued a Special Law in Force for pine wilt disease control. The Special Law continued the national project for reducing pine losses and stopping disease spread. Aerial spraying of organophosphate insecticides on a large scale was one of the major control methods in this national project. Over a 10-year period, an average of 123 000 ha of pine forests, equivalent to 20% of the total infested area, was sprayed annually. In later years, however, aerial sprayed areas were reduced to less than 50 000 ha.

The annual budget in 1986 was 6 billion yen (*ca* US $50 million), 55% of which was for aerial spraying, 8% for ground spraying, and the rest for eradicating dead trees by felling followed by chemical treatment such as spraying insecticides or fumigating and chipping or burning the dead logs. After losing the power of the Special Law in 1997, the government continues its control efforts based on the common law, such as laws for forest protection. Government policy to control the disease puts emphasis on pine forests closely associated with public benefit. Priority control is focused on pine forests which play a role in erosion control, water conservation, sand stabilisation, wind protection, and in maintaining aesthetic value. In heavily infested areas, conversion of pine forests to other tree species is encouraged in order to protect pine forests of public benefit and to maintain the function of forests in land conservation. Another purpose of forest conversion is to protect pine forests of public benefit by removing the surrounding pine forests which are potential infection sources. Forests consisting of other tree species serve as barrier zones for the disease. Control efforts are also focused on pine forests now at a primary stage of infestation in the northern

Table 2. *Resistance and susceptibility of* Pinus *species to the pine wood nematode.*

Origin	*Pinus* taxon	Age (y)	Tree survival	
			(n)	(%)
Asia	*P. armandi*	37	7	38.9
	P. fenzeliana	9	8	100
	P. griffithii	20	108	83.1
	P. koraiensis	22	134	28.5
	P. morrisonicola	27	10	100
	P. huangshanensis	8	15	100
	P. massoniana	34	21	27.6
	"	23	25	67.6
	P. tabulaeformis	35	3	60
	P. khasya	19	0	0
North America (East)	*P. strobus*	33	60	96.8
	P. clausa	19	21	100
	P. echinata	39	22	100
	"	18	140	100
	P. elliottii	31	450	100
	P. palustris	37	61	100
	P. taeda	34	350	100
	P. virginiana	39	24	100
	P. pungens	34	24	100
	P. rigida	39	29	93.5
	P. banksiana	37	9	69.2
North America (West)	*P. monticola*	25	4	80
	P. contorta	19	9	33.3
	P. jeffreyi	36	4	36.4
	P. muricata	41	0	0
	P. radiata	40	24	20.4
	P. ponderosa	39	22	46.8
North America (Southwest)	*P. leiophylla*	20	59	54.6
	P. engermannii	20	137	79.2
	P. ayacahuite	37	5	41.7
	P. cooperi	26	2	16.7
	P. durangensis	26	3	33.3
	P. gregii	26	16	72.7
	P. hartwegii	26	1	20

Table 2. *(Continued).*

Origin	*Pinus* taxon	Age (y)	Tree survival	
			(n)	(%)
	P. patula	39	9	81.8
	"	26	2	20
	P. michoacana	21	4	22.2
	P. oocarpa	21	2	5.1
	P. pseudostrobus	21	3	18.8
Europe	*P. nigra*	23	6	6
	"	20	13	28.3
	P. peuce	38	17	94.4
	P. pinaster	40	10	14.1
	"	35	25	12
	"	22	104	47.3
	P. pinea	39	3	23.1
	P. sylvestris	37	2	7.1

(After Furuno *et al.*, 1993.)

regions and inland mountainous areas where infestation centres have only recently been established.

The final resolution for controlling plant disease caused by exotic pathogens should rely on either complete eradication of the pathogens or resistance breeding strategies. Throughout the national project for breeding resistant clones of native pines, 92 clones of *P. densiflora* and 16 clones of *P. thunbergii* were selected, and seed orchards for those resistant clones were established. In 1999, more than 500 000 seedlings of resistant clones were supplied for planting.

Regarding the resistance and susceptibility to pine wilt disease of various pine species distributed worldwide, there is literature with the results of surveys, even though surveys were carried out of trees under natural conditions in Japan (Furuno *et al.*, 1993). Survival rates of 52 exotic pine species planted in Kyoto University forests in Kyoto and Wakayama were investigated throughout the 1960s and 1990s and were recorded. Most pine trees were more than 20 years old and grew under conditions of possible exposure to natural infection by the pine wood nematode. *Pinus densiflora* and *P. thunbergii* grown at the same localities were heavily affected by the disease during those periods and no trees survived. Results are compiled in Table 2 based on the original data. It was clearly indicated that almost all pine species native to eastern North America were highly resistant and most pine species native to Europe were highly susceptible to the disease. Survey results on the resistance of many other pine species distributed worldwide will help to estimate the possibility of disease spread in various regions throughout the world.

Acknowledgement

The author kindly thanks Dr J.W. Webster, Simon Fraser University, for reviewing the manuscript.

References

FURUNO, T., NAKAI, I., UENAKA, K. & HAYA, K. (1993). [The pine wilt upon the exotic pine species introduced in Kamigamo and Shirahama Experiment Station of Kyoto University – Various resistances among genus *Pinus* to pine wood nematode, *Bursaphelenchus xylophilus* Steiner and Buhrer.] *Report of the Kyoto University Forests* 25, 20-34.

KIYOHARA, T. (1973). [Effect of temperatures on the disease incidence of pine seedlings inoculated with *Bursaphelenchus lignicolus.*] *Transactions of the 84th Meeting of Japanese Forest Society*, pp. 334-335.

KIYOHARA, T. & TOKUSHIGE, Y. (1971). [Inoculation experiments of a nematode, *Bursaphelenchus* sp., onto pine trees.] *Journal of Japanese Forest Society* 53, 210-218.

MAMIYA, Y. (1988). History of pine wilt disease in Japan. *Journal of Nematology* 20, 219-226.

MAMIYA, Y. & ENDA N. (1972). Transmission of *Bursaphelenchus lignicolus* (Nematoda: Aphelenchoididae) by *Monochamus alternatus* (Coleoptera: Cerambycidae). *Nematologica* 18, 159-162.

MAMIYA, Y., KOBAYASHI, T., ZINNO, Y., ENDA, N. & SASAKI, K. (1973). [Disease development of pine trees naturally infected with *Bursaphelenchus lignicolus.*] *Transactions of 84th Meeting of Japanese Forest Society*, pp. 332-334.

MORIMOTO, K. & IWASAKI, A. (1972). [Role of *Monochamus alternatus* (Coleoptera: Cerambycidae) as a vector of *Bursaphelenchus lignicolus* (Nematoda: Aphelenchoididae).] *Journal of Japanese Forest Society* 54, 177-183.

YANO, M. (1913). [Investigation on the cause of pine mortality in Nagasaki Prefecture.] *Sanrinkoho* 4 (Suppl.) 1-14.
ZINNO, Y., TAKIZAWA, Y. & SATO, H. (1987). [Pine wilt disease and its control in cool areas.] *Series of Manual of Forest Research Sciences* No. 86, 75 pp.

Nematology Monographs & Perspectives, 2003, Vol. 1, 21-24

The history, dispersal and potential threat of pine wood nematode in China

Bao Jun YANG

The Chinese Academy of Forestry, Beijing, China 100091

Summary – Since the pine wood nematode (PWN) was discovered in Jiangsu Province in 1982, it has occurred in seven provinces in China. The long distance spread of PWN is *via* the transport of infected logs and products made from infected pine wood from other countries or other places inside China where pine wood nematode is present. PWN is an enormous potential threat to pine forests in South and Central China, because the temperature in this area is favourable, the vector beetle is present, the pine species are susceptible and PWN already occurs in some places. Pines have been very important forestry trees in China. The area of planted pines is about 20% of total forestry. Pines are considered as the symbol of longevity for the Chinese people, who are consequently very fond of them. Since the pine wood nematode was discovered at Nanjing City, Jiangsu Province, this nematode has occurred in 59 counties (or cities) in six provinces. The federal and local governments have paid great attention to this disease, because pine wood nematode has destroyed about 20 million pine trees in 75 000 ha, and this nematode is spreading further.

History

Pine wood nematode was first discovered in 1982 in dead Japanese black pine, *Pinus thunbergii* at Nanjing City, Jiangsu Province. This nematode now occurs in seven cities. The pine species damaged by pine wood nematode are Japanese black pine and masson pine, *P. massoniana*, as well as cluster pine, *P. pinaster*, and Japanese red pine, *P. densiflora*, in this province, but there are fewer dead trees among the latter two pine species. The control methods used by local people are cutting dead pines, treating the logs of dead pines (usually burning, fumigation by methyl bromide, or chipping), and spraying chemicals (generally fenitrothion) to control beetles. However, the nematode continues to spread. In 1988 pine wood nematode occurred in Anhui Province, a neighbouring province to Jiangsu. The situation of

Anhui is the same as that found in Jiangsu. The pine species infected by pine wood nematode are masson and Japanese black pine. In the same year the nematode spread to Shenzhen City, Guangdong Province, from Hong Kong. Pine wood nematode has now already been found in four cities. Control techniques widely used by local people are felling dead pines and treating them, spraying chemicals, using attractants and a species of bethylid, *Scleroderma guani*, to control beetles. Results are quite good. The number of wilted pines was reduced by 98% in 1999-2000. The species infected by pine wood nematode in this province is masson pine.

In 1991, pine wood nematode occurred in Zhejiang Province, neighbouring province to Jiangsu and Anhui Provinces. The species infected here by pine wood nematode are masson and Japanese black pine. The disease is more widespread in Zhejiang Province. In the same year, the nematode was discovered in Changdao County, Shandong Province. This area is an island that was formerly a harbour where many logs were imported from North America. Fourteen dead pines were found and cut in the first year, but pine wood nematode has not been eliminated. In 1998 we used early diagnosis (to ascertain eleoresin levels) to diagnose about 0.6 million pines, felled those showing abnormal eleoresin levels and burned them. Dead pines were reduced by 97% in 2000. The diseased species are Japanese black and Japanese red pines. The pine wood nematode was discovered in masson pine in Hubei Province in 1999. The local people cut infected and weak trees; however, about 200 pines were still infected in 2000.

The insect vectors of pine wood nematode were investigated. The most important effective vector is *Monochamus alternatus*. This insect is widely distributed throughout China and occurs in almost all the planted pine forests in central and southern China.

Dispersal

The natural spread of pine wood nematode by insect vector is not very important, because the distance travelled by the beetles is limited. The long distance spread of pine wood nematode is *via* the transport of infected logs and products made of infected pine wood from other countries where pine wood nematode is present, or from diseased places inside China.

Some phenomena show that the pine wood nematode may be from other countries. For example, the pine wood nematode was first reported in Zhongshanling Park, Nanjing City. However, some local people told me that pine wood nematode first occurred in a large hotel behind Zhongshanling Park, perhaps in electrical equipment imported from Japan. To date at least 15 quarantine bureaux have found pine wood nematode in packing boxes or stay wood.

The transport of infected logs or products, such as packing boxes, wire reels, and even furniture, spreads the nematode. For example, on Fangshan tree farm in Nanjing City, a farmer took a log from a diseased place for building a pigpen, and pines were subsequently infected around his house. So preventing the transport of logs and products made from pine trees infected by pine wood nematode is very important to control the spread of this nematode.

The potential threat of pine wood nematode to China

Four factors determine the occurrence and epidemic of pine wilt disease caused by pine wood nematode. They are the susceptible hosts, effective insect vectors, suitable environmental conditions and the pine wood nematode. If the four factors exist simultaneously the disease can occur and spread fast.

The main pine species in China are Korean pine, *Pinus koraiensis*; Mongolian pine, *P. sylvestris* var. *mongolica* in the north-east; Chinese pine, *P. tabulaeformis*; lacebark pine, *P. bungeana* in central north; armand pine, *P. armandi*, in the west; Japanese black pine, *P. thunbergii*; Japanese red pine, *P. densiflora*, in the east; masson pine, *P. massoniana* in the south; Himalayan pine, *P. griffithii* in Yunnan and Tibet; Huangshan pine, *P. taiwanensis* in the south-east; Yunnan pine, *P. yunnanensis* in the south-west; Kwangtung pine, *P. kwang-tungensis* in Guangdong, Guangxi and Guizhou, and fenzel pine, *P. fenzeliana*, in Hainan, Guizhou and Guangxi. Almost all pine species are susceptible to varying degrees.

In China the most important vector of pine wood nematode is *Monochamus alternatus*. This beetle is found in most provinces in central and southern China, including Hebei, Shanxi, Shaanxi, Sichuan, Shandong, Henan, Anhui, Jiangsu, Hubei, Tibet, Yunan, Guizhou, Jiangxi, Fujian, Guangdong, Guangxi, Hainan, Beijing, Shanghai, Hong

Kong, as well as Taiwan. In the north-east, including Heilongjiang, Jilin, Shandong and Hebei, another two vector insects, *Monochamus saltuarius* and *M. sutor*, exist.

The most important environmental factor influencing the occurrence and spread of pine wilt disease is temperature. In Japan, pine wood nematodes can commonly occur and damage pine forest seriously in areas where the mean annual temperature is above 14°C, and can occur in, but not damage seriously, pine forest in areas where the mean annual temperature is 10°-14°C. In China, the mean annual temperature is above 14°C in the large area south of Jiangsu, Anhui, Henan, south Shaanxi, and central Sichuan. In central and southern China, the temperature is suitable for pine wilt disease.

Pine wood nematode is a great potential threat to pine forests in southern and central China. Because the temperature here is favourable for the disease, the vector beetle exists widely, and the pine species are susceptible to varying degrees. Meanwhile, pine wood nematode already occurs in some places. Hence it is most important to stop the invasion of pine wood nematode into virgin areas, and if pine wood nematode appears in new areas, we should take emergency measures to eradicate it as soon as possible.

Nematology Monographs & Perspectives, 2003, Vol. 1, 25-30

Pine wilt disease – a threat to pine forests in Europe

Kazuo SUZUKI

The University of Tokyo, Tokyo 113-8657, Japan

Summary – Pine wilt disease, which is endemic in North America, has already spread epidemically to East Asia, including Japan, China, Taiwan and Korea, and now to Portugal in Europe. If it were to become established in the pine forests of Europe, it could become one of the most serious threats to pine forests worldwide in the 21st century.

The 21st century is an environmental age. 'Green' will be the key word all over the world. People's feelings towards forests and trees are stronger than ever before. In 1997, in the world famous scientific journal *Nature*, the economic value of forests was estimated to be worth 4700 billion US$ per year, 14% of total global flow value. In Japan, the most beautiful scenery is built up along the seaside by pine trees. These pine forests actually make up 10% of the forested land that covers 67% of Japan. Pine forests play an important role for both society and the environment. The function of forests is not limited to timber production. The value of the public function of Japan's forests was estimated initially to provide 12 trillion yen per year in 1972, then, depending on a rise in market price, 39 trillion yen in 1991, increasing to 75 trillion yen in 2000. Since Japan's national budget is approximately 70 trillion yen, the invisible value created by Japan's forests is huge. The Science Council of Japan is currently discussing the correct way, from a scientific viewpoint, to evaluate the value of forests. In recent years, the decline of forests worldwide has been seen as a problem. In the Northern Hemisphere, the major forests are composed of pines. Pines have spread all over the Northern Hemisphere since their incipience in the Mesozoic era. Their original birthplace was supposed to be in the Bering Sea where the land was connected with Alaska and Siberia. At present, more than 100 species are recognised and distributed from far north to tropical latitudes. The tallest pines in the world are the sugar pines, *Pinus lambertiana*, reaching 70 m in height in Sierra Nevada, and the oldest is the bristlecone pine, *Pinus longeava* (*P. aristata*), in Nevada and California, *ca* 5000 years old, or at least 4844 years from the study of its rings. Thus, throughout history, man has shown a special affinity for pine trees. The Loess Plateau of China, the middle regions of the Yellow River basin in northwestern China, where we have been working on a 'Rehabilitation of the Loess Plateau' project, covers an area roughly one and a half times that of Japan and has a population of 90 million people. The fields in this region are terraced and all fields and gullies are covered with crops such as corn, millet, soybean and sunflowers. It is said that cultivation of terraces leads to Heaven. Responding to growing population pressure, the land in this region is being extensively used for grazing goats and sheep. It is feared that erosion in the V-shaped valleys caused by this grazing will accelerate the process of desertification. In this region today, forests cover no more than 6% of the Loess Plateau. However, it is said that the entire plateau was very fertile and thickly forested until 3000 years ago. We are trying to reforest this area by selecting forest trees such as *Pinus sylvestris* var. *mongolica* and *P. tablaeformis*. Thus, forest ecosystems are indispensable to human activities. Professor Nicholas T. Mirov, University of California, Berkeley, said in his book 'There are other useful conifers – spruce, Douglas fir, larch, hemlock – but none are as outstanding in importance as pines. Pines have affected man's religion, his dwellings, and his livelihood'. Plant diseases first became a concern to the public in the mid-19th century. As to forest diseases, three major epidemic diseases happened in the 20th century, such as Dutch elm disease, chestnut blight, and white pine blister rust that spread around the world. First, Dutch elm disease broke out in Holland at the beginning of the 20th century. The money spent on its control has now reached US$15 billion. Second, the chestnut blight caused almost complete destruction of American chestnut by the introduction of pathogens from East Asia, probably Japan. In those days American chestnut was the main timber in the American East. Third, white pine blister rust spread from Far East Asia. Then the most serious epidemic disease – pine wilt disease – broke out in the late 20th century.

Since the end of World War II in 1945, pine wilt disease has caused the loss of 26 million m^3 of timber. This volume is equivalent to timber required for the construction of 1.7 million wooden houses in Japan. Initially, the disease was recognised as an unusual wilt of pines in 1905 in the port city of Nagasaki located in southern Japan, which was the only port open to foreign trade in the 19th century in Japan.

Pine wilt disease spread from many port cities in western parts of Japan thereafter. Then, the damage increased to 1 million m^3 per year in 1947. In 1950, in order to control the disease, recommendations were made to develop a pine bark beetle control programme to control 'Matsukuimushi', a general term covering more than 70 species of pine bark and wood borers which were thought to be concerned with pine wilt disease. The overall recommendations were the direct control of 'Matsukuimushi' by felling, de-barking, and burning. These measures immediately proved effective against the disease; however, in practice, they required a huge input of manpower. With the high level of economic growth in the 1960s, felling and burning were replaced by spraying of lindane (γ-BHC) insecticide. Since 1971, when lindane was banned, other organophosphate insecticides with a shorter residual activity, such as fenitorothion (sumithion), have been used. Owing to the recommendations, pine timber losses fell to 0.2-0.3 million m^3 annually. However, pine timber losses again increased from the late 1970s, reaching a maximum of 2.4 million m^3 in 1979.

Up until now, more than 2000 papers have been published on pine wilt disease. The causal agent of the disease, pine wood nematode (PWN), *Bursaphelenchus xylophilus*, was recovered from dead Japanese black pines in 1969, and was determined as the causal agent of pine wilt disease by the application of Koch's postulates in 1971. The susceptible pine species to pine wilt disease are Japanese black pine (*Pinus thunbergii*), Japanese red pine (*P. densiflora*) and *P. luchuensis*, which are also the most predominant pine species in Japan.

In 1979, PWN was discovered in Missouri, USA. This was the first report of PWN in USA published in the *Journal of Plant Disease*. Taking this opportunity, the USA-Japan seminar 'The resistance mechanisms of pines against pine wilt disease' was held in Hawaii in 1984 in coordination with Professors Dropkin and Oku. One year later another symposium entitled 'Etiology of pine wilt and vector relationships of *Bursaphelenchus xylophilus*' was held in conjunction with the annual meeting of the American Phytopathological Society in Reno, Nevada, in 1985. Thereafter, the APS Symposium series 'Pathogenicity of the pine wood nematode' was edited and published by Professor Michael J. Wingfield. At present, most pine species in the USA seem to be moderately resistant to PWN and PWN was determined as being endemic to the USA.

In recent years, pine wilt disease has spread to Nanjing, China, in 1982 on Japanese black pine and southern red pine (*P. massoniana*), and to Taiwan in 1985 on *P. luchuensis* and Japanese red pine, and to Pusan, Korea in 1988 on Japanese black pine and Japanese red pine. As a result of the global concern on pine decline, an international symposium on 'Sustainability of pine forests in relation to pine wilt and decline' was held in Tokyo, Japan, in 1998 as an all Division 7 meeting of IUFRO. We took this opportunity to build up a new Working Party in IUFRO, that is, 7.02.10. 'Pine Wilt Disease' with the coordinator Professor Marc J. Linit and deputy coordinators, Christer Magnusson and Kazuyoshi Futai.

Furthermore, quite recently PWN was found in a dead maritime pine (*P. pinaster*) in Portugal in 1999. This was the first report of the occurrence of PWN in Europe.

Up to now, PWN has been found in many pine species in Asia and North America. In addition to these pine species, PWN is also naturally found in both diseased blue spruce and white spruce (*Picea* spp.), Eastern larch and European larch (*Larix* spp.), Douglas fir (*Pseudotsuga* spp.), balsam fir (*Abies* spp.), and Atlas and Deodar cedars (*Cedrus* spp.) in North America.

Wilting mechanism and environmental conditions

PWN introduced into the shoots of young pine trees during maturation feeding of Japanese pine sawyer migrate rapidly into the whole trunk of pine trees. After the entry of PWN into living pines, a slight reduction in the flow of oleoresin exudate is observed as a unique symptom at early stages of the disease. This is due to the movement of PWN through resin canals at early stage of the disease development and epithelial cells around resin canals are destroyed. Nematodes then eventually move from rays to tracheids through pits. At the same time, enhanced ethylene production is observed 2-3 days after invasion by PWN. While the exact mechanism remains unclear, this increase is incited by the excretion of a considerable amount of cellulase by the pine wood nematodes. The nematode density is very low at an early stage of disease development, often as low as a few nematodes per 100 g fresh weight of wood, even following highly concentrated inoculations.

As a general rule, symptom development is divided into two stages following invasion of the wood of pines by nematodes, namely an early stage and an advanced stage. In the early stage, cytological changes in the xylem parenchymatous cells occur, and these are soon followed by cavitation and embolism formation within a number of tracheids. Dysfunction of conduction in the vascular system has been shown by acid fuchsin absorbed from the roots. Such internal symptomatology is induced not only in compatible but also in incompatible combinations of pine tree and nematode isolates. However, growth of the nematode population in living pines under conditions unfavourable to the nematode is not assured, even if a high concentration of nematodes is inoculated. Therefore, this stage is considered to be latent, that is, denaturation of parenchymatous cells by nematode invasion results in cavitation and embolism of some tracheids.

At the onset of the advanced stage of disease development, visible symptoms are expressed as a severe reduction of the oleoresin exudation rate and chlorosis of 2-3-year-old needles, accompanied by a decrease in transpiration. This phenomenon is a unique characteristic of pine wilt disease, accompanied by a further increase in ethylene production. Furthermore, cambial death and cavitation occur within a large part of the outer xylem, and result in a water deficiency which induces a decrease in both transpiration and photosynthesis in leaves. At this time, other pathophysiological phenomena are observed; for example, electrolyte leakage from pine tissues occurs, and a number of abnormal metabolites such as benzoic acid are produced. From the onset of water stress, the nematode population begins to increase in remarkably close correlation with time.

In terms of disease development, the water status of pines plays a very important role in the pine-nematode relationship. Experimental results suggest that pine seedlings do not wilt solely by virtue of the number of nematodes existing under conditions unfavourable to them, such as a well-watered environment. Empirically, pine wilt disease seems to occur more frequently and to be more destructive in summers with little rainfall. Therefore, these two factors, the physiological water status and a particular nematode population density, are considered to be decisive factors in disease development.

Pine wilt disease control programme

Before the discovery of pine wood nematode as the causal agent of pine wilt disease, nobody knew the actual pathogen of pine wilt disease. However, felling, de-barking and burning of damaged pines immediately proved effective against the disease. Following the discovery of the pine wood nematode as the pathogen of pine wilt disease, the Japanese pine sawyer, *Monochamus alternatus*, was determined as a vector of the pine wood nematode. Accordingly, control measures were redirected to the destruction of the Japanese pine sawyer by the spraying of insecticide.

A large scale, 5-year control project for pine wilt disease was initiated in 1977 by the Special Law in Force. It was characterised by implementation of aerial spraying to prevent Japanese pine sawyer maturation feeding. In spite of many efforts, severe damage was not completely controlled and the disease spread widely from the southern to the northern part of Japan. The reasons for complete control failure are thought to be: *i*) the limitations of aerial spraying; *ii*) too much reliance on the Special Law in Force; and *iii*) those people concerned not recognising the severity of the threat to pine forests. In 1997, the Special Law in Force came to an end after three revisions and enforcement for 20 years since 1977. Currently, annual timber losses still amount to less than 1 million m^3 annually. The pine wilt disease is still a great threat to both community and landscape in Japan.

Present pine wilt control programme and threat of epidemics

Control measures against pine wilt disease are aimed at breaking the pine tree-pine wood nematodes-pine sawyer disease triangle. Current control measures consist, for the most part, of aerial spraying of insecticides effective against pine sawyer as a disease preventative to cut pine tree-pine sawyer relations, spraying of insecticides on timber damaged by infestation to kill the pine sawyer, and trunk injection of chemicals active against pine wood nematode to cut pine tree-PWN relations. In spite of various efforts, the total amount of pine timber lost to the disease is not decreasing conspicuously.

The incidence of pine wilt disease is closely related to environmental conditions as mentioned previously. Environmental shifts such as increasingly warm and unusual weather conditions, which are expected in the near future, most certainly affect susceptibility of pine trees. In addition, shifts in forest ecosystems resulting from the deposition of acidifying substances may significantly influence pine wilt disease. Furthermore, pine wilt disease has the potential to become a major threat if it were to be exported to European countries, because Scots pine (*P. sylvestris*) and maritime pine (*P. pinaster*) are very susceptible to this disease.

Nematology Monographs & Perspectives, 2003, Vol. 1, 31-54

Pine wood nematode: pathogenic or political?

Robert I. BOLLA and Robert WOOD

Department of Biology, Saint Louis University, 3507 Laclede, St. Louis, MO 63103-2010, USA

Summary – The European Union has banned shipment of untreated pine wood from North America and several Asian countries into Europe because of the potential for introduction of members of the pine wood nematode species complex, *Bursaphelenchus* spp. Some species of this nematode are pathogenic in some conifer species under particular environmental conditions and, as a consequence, have caused extensive damage to pine forests in Japan, China and Korea. The import ban is based on the lack of identification of *Bursaphelenchus xylophilus*, the species considered to be most pathogenic, in Europe until the recent identification in Portugal. It is felt that, if introduced, the nematode would establish populations or interbreed with endemic non-virulent species to result in a situation similar to that seen since the early 1970s in Japan. This ban has had major consequences on the North American forest industry. Recently many new species of *Bursaphelenchus* have been described from dead or dying pines throughout Europe. These identifications have been based on morphological characters that are limited in usefulness for species descriptions and cannot be used to differentiate populations. Thus other taxonomic characters, including molecular taxonomy, have become important. We will look at the accuracy of methods used for species identification and at what criteria might be used to define and differentiate species of *Bursaphelenchus* when considering import and export bans.

Pine wood nematode (PWN, *Bursaphelenchus xylophilus*) is one species of *Bursaphelenchus* associated with dead or dying pines or hardwoods (Steiner & Buhrer, 1934; Baujard, 1980; Webster *et al.*, 1990; Beckenbach *et al.*, 1992). *Bursaphelenchus xylophilus* is considered to be virulent, whereas other species, including *B. mucronatus*, are considered non-virulent (Kondo *et al.*, 1982; Wingfield *et al.*, 1983; Mamiya, 1984; Dwinell, 1985, 1987; Bolla *et al.*, 1986; Bedker *et al.*, 1987; Dwinell & Nickle 1989; Kiyohara & Bolla, 1990; Evans *et al.*, 1996; Fielding & Evans; 1996; Bolla & Wood, 1999). Under the right conditions, however, *B. mucronatus* may be virulent in pines (Rutherford & Webster, 1987; Rutherford *et al.*, 1990, 1992). Other species recovered from dead or dying trees, including *B. fraudulentus* from *Prunus* spp. and *B. hoffmani*, have not been tested rigorously following Koch's postulates for virulence (Rühm, 1956; Skarmoutsos *et al.*, 1988; Yin *et al.*, 1988; Schauer-Blume & Sturhan, 1989; Kulinich *et al.*, 1994; Philis, 1996; Philis & Braasch, 1996; Braasch, 1998, 2001; Braasch *et al.*, 1998; Caroppo *et al.*, 1998; Kanzaki *et al.*, 2000; Skarmoutsos & Michalopoulos-Skarmoutsos, 2000). Mamiya and Enda (1979) demonstrated unequivocally that the original isolate of *B. xylophilus* isolated from Japanese black pine (*Pinus thunbergii*) was virulent on immature and adult pines in greenhouse, screenhouse, and field. Kondo and co-workers (1982) and Dropkin and Foudin (1979) used similar experiments to demonstrate the virulence of *B. xylophilus* (isolate US12) from Missouri. This established the first evidence of a virulent strain of PWN in North America. Pathogenesis of *B. xylophilus* depends on origin of the population, host species, and environmental factors. The most important environmental factors are the mid-summer isotherm and water status of the pine relative to the potential for cavitations of the xylem and phloem (Suzaki & Kiyohara, 1978; Rutherford & Webster, 1987; Ikeda *et al.*, 1990; Kuroda, 1991; Ikeda & Kiyohara, 1995; Ikeda, 1996a, b; Ishida & Hogetsu, 1997; Rutherford *et al.*, 1990, 1992; Jikumaru & Togashi, 2000). PWN is most pathogenic on three needle pines planted where the midsummer isotherm is greater than 20°C and where water stress is a common seasonal event (Rutherford & Webster, 1987; Rutherford *et al.*, 1990, 1992). This is not unexpected as the mechanism of pathogenesis is, at least in part, due to cavitations of the xylem and phloem when soil water potential is low and the pressure of the upward water stream decreases during periods of high rates of transpiration (Ikeda *et al.*, 1990; Ikeda & Kiyohara, 1995; Ikeda, 1996a, b; Ishida & Hogetsu, 1997; Kuroda, 1991; Jikumaru & Togashi, 2000). It might be proposed than any species of *Bursaphelenchus* which disrupts the integrity of epithelial cells of the resin canals such that resin can leak into tissue where it is exposed to molecular oxygen for oxidation, is pathogenic.

Bursaphelenchus xylophilus is endemic to North America, on pines planted where the midsummer isotherm is greater than 20°C (Rutherford *et al.*, 1992). Thus PWN in North America is a pathogen mainly in horticultural plantings; otherwise the nematode and pine coexist

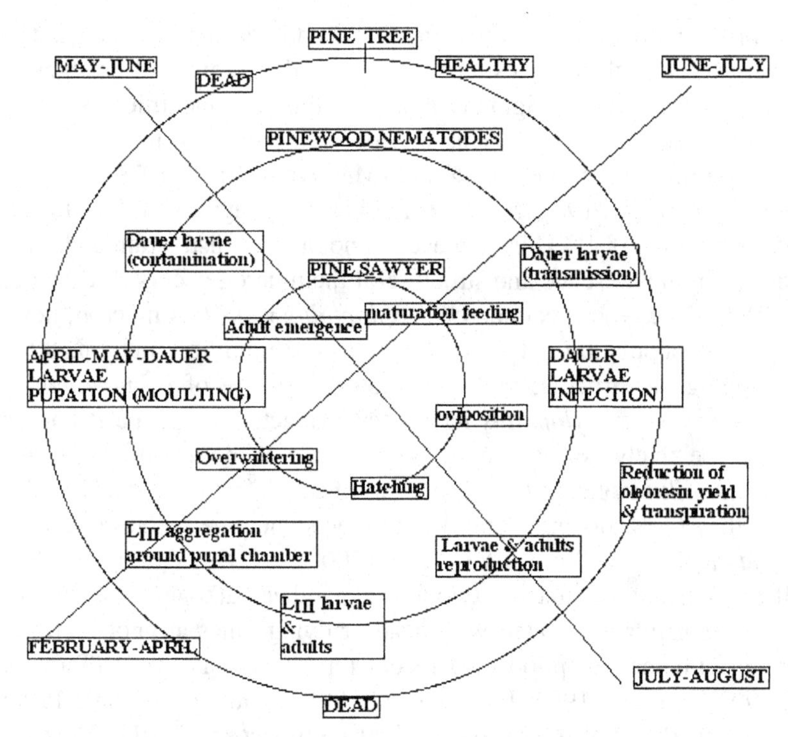

Fig. 1. *Life cycle of* Bursaphelenchus xylophilus *in Japan and North America showing the seasonal effects of population development and differentiating between the life cycle in living and in dead pine. Modified from Mamiya (1986).*

suggesting co-evolution. As an introduced species in Japan, however, *B. xylophilus* is responsible for widespread damage to pine forests (Mamiya, 1984, 1988). Similarly, PWN has caused extensive loss of pines in China and Korea following its introduction through shipping (Choi & Moon, 1989; Yang *et al.*, 2000).

The life cycle of PWN is the final brush stroke in painting pine wood nematode as a potential pathogen (Fig. 1). Members of the genus *Bursaphelenchus* are generally mycophagus phoretic with insects feeding on fungi associated with the insect. This is true of *B. xylophilus* which, if transmitted to dead pines during egg-laying by the vector insect, may complete its life cycle feeding on fungi in dead pines with no need to include living pines in the life-cycle (Yoshimura *et al.*, 1999; Aikawa & Togashi, 2000; Anbutsu & Togashi, 2000).

Other species of *Bursaphelenchus* associated with dead or dying pines have been differentially characterised as to virulence. This is based, however, primarily on infectivity to immature seedlings, and thus the validity of these tests might be questioned based on developmental changes in the pine as it matures. Originally *B. mucronatus* was identified as a non-virulent species from Japan. This species is differentiated from *B. xylophilus* by a mucron on the female tail (Mamiya & Enda, 1979; Baujard, 1980; Kondo *et al.*, 1982; Mamiya, 1984, 1988; Magnusson & Kulinich, 1996; Braasch, 2001). This, however, is not a clear distinction and wide variation in the presence and structure of the mucron occurs (Webster *et al.*, 1990). The type specimen of *B. xylophilus* lacks this mucron; however, other populations of this species have a modified mucron, often overlapping the structure of the mucron descriptive of *B. mucronatus*. These forms of *B. xylophilus* have been characterised as the R and M form, respectively (Webster *et al.*, 1990). Two of the reported M forms of *B. xylophilus* from North America have been defined as *B. mucronatus* where they show moderate pathogenesis on some pine species (Beckenbach *et al.*, 1992, 1999; Bolla & Wood, 2000).

It has been stated that *B. xylophilus* or other pathogenic species of *Bursaphelenchus* associated with dead or dying pines are not found in Europe with the exception of a recent report from Portugal (Tares *et al.*, 1992; Braasch, 1994; Evans *et al.*, 1996; Fielding & Evans, 1996; Braasch *et al.*, 1999; Mota *et al.*, 1999; Sousa *et al.*, 2001). Thus the potential for populations introduced from North America and Japan to cause widespread devastation of pine has been postulated (Evans *et al.*, 1996). This has led to a ban on import of non-treated pine wood products from North America with a consequent economic impact (Evans *et al.*, 1993).

There has been little overall effect on export of pine from the USA (Fig. 2) as new markets have been found and export of different species of coniferous wood has increased. However, if one looks at the export of individual pine species such as yellow pine (Fig. 3) it is clear that the European ban has had an impact. The effect on Canadian forestry is considerably more significant (Webster, pers. comm.).

The purpose of this paper is to look at the methods used to describe relationships between isolates and species of *Bursaphelenchus* found associated with dead or dying conifers, and to discuss the criteria of decision-making to evoke import bans. A variety of criteria have been used to explore relationships and a bioinformatics approach has been used to evaluate available data to describe the PWNSC (Pine Wood Nematode Species Complex). I will discuss the benefits and problems of

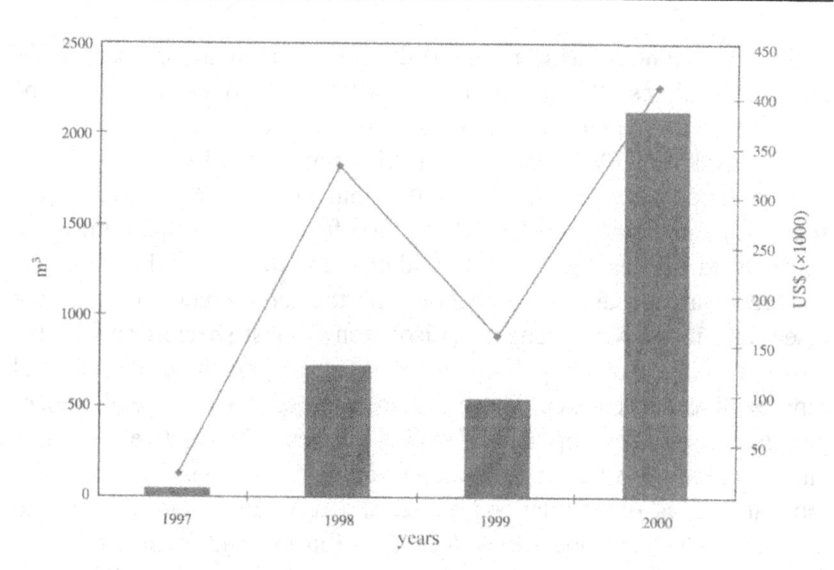

Fig. 2. *Exports of chipped or sawn pine wood from the USA for years 1997-2000. Bars indicate volume (m^3) of wood exported; the solid line represents the value of the exports in US dollars.*

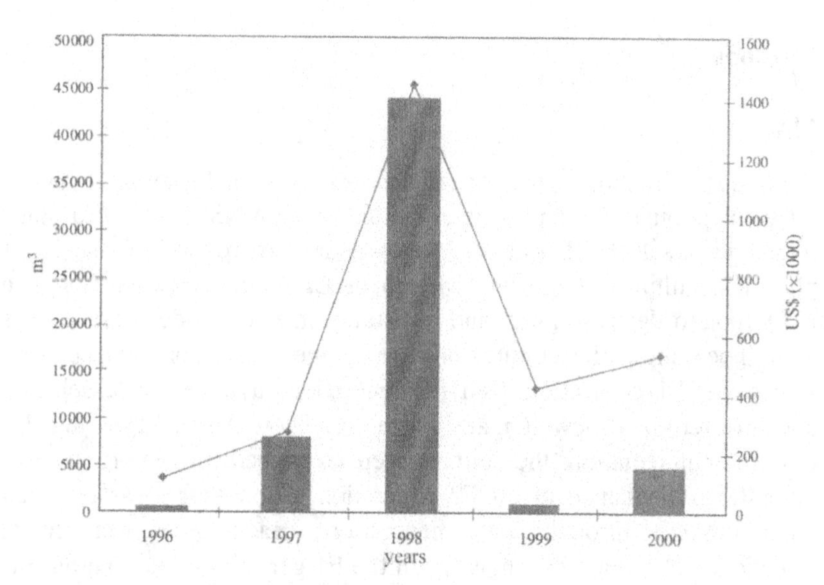

Fig. 3. *Export of yellow pine from the USA to Europe for the years 1996-2000. The solid bars indicate volume (m^3) of kiln dried yellow pine lumber exported; the solid line represents the value in US dollars.*

molecular taxonomy as a means of describing populations and species and I will address the question as to whether major decisions can be made based on sequence differences in a single gene.

In considering PWN and relationships within the PWNSC (defined here as a complex of species and populations of *Bursaphelenchus* spp. associated with dead or dying conifers or found phoretic with cerambycid beetles), we must consider: *i*) the duality of the life cycle between parasitic and mycophagous; *ii*) the requirement for a vector insect and its dispersal range; *iii*) isolation of host species; and *iv*) the effect of isolation of a population within a tree on gene flow and potential reproductive bottlenecks. These parameters could lead to isolation of populations and development of sibling species, reproductively isolated and differentially virulent compared to other populations. We must also look at origins of populations expressing different pathogenicity and ascertain whether populations found in Europe and Asia are native populations or if they are introduced. Finally, species relationships must be considered, particularly how they might relate to differentiation of the R and M forms of *B. xylophilus*. What suite of criteria then would be acceptable in describing new species of *Bursaphelenchus*?

Methods

MATING

Single pair matings were done to determine mating behaviour between different populations and species within the PWNSC. A fourth stage female (J4), selected based on development of reproductive tissue, was placed in culture on *Botrytis cinerea* or *Ceratocystis ips* growing on PDA (potato dextrose agar) and a mature male was added. Three days after inoculation, the cultures were examined daily for presence and size of an F1 generation. Ten F1 generation juveniles were selected and transferred to new fungal cultures on PDA. Again beginning 3-5 days after inoculation, the cultures were examined for the presence of juveniles as evidence of an F2 generation. Sub-culture was repeated when possible through seven generations. Backcrosses were made selecting a J4 female or a male from the F1 generation and mating this with a male or J4 female from the parental stock. Matings were scored as positive only when cultures could be maintained for at least seven generations and backcrosses were positive.

CHROMOSOME COMPLEMENT

Nematodes were frozen on dry ice and squashed as described by Bolla and Boschert (1993). The nematode squashes were stained as described by Bolla and Boschert (1993). Haploid chromosome number in testes or ovary was scored.

MOLECULAR TAXONOMY

Populations of *Bursaphelenchus* (Table 1) were raised *in vitro* on *Botrytis cinerea* (Bolla *et al.*, 1986) with passage once a month through 6-8-year-old field-raised pines to maintain virulence and 'wild type' characters. Isolates were used to evaluate sequence similarities or differences in the DNA sequence within the internal transcribed spacer (ITS) region of ribosomal DNA or in Hsp 70 (heat shock protein 70s) DNA. Five to ten adult nematodes were hand selected from each population and DNA extracted from individual nematodes. Nematodes were frozen immediately in liquid nitrogen and immersed in nematode buffer (5 mM Tris HCl, pH 8.0; 2.5 mM NaCl; 0.1 mM EDTA; 2% SDS) and 1 mg of Proteinase K was added. The mixture was heated for 1 h at 65°C. DNA was then extracted (Maniatis *et al.*, 1989). The ITS regions were amplified using the primers 5′ACCTTGTTACGACTT3′ and 5′CTTTGTACACACCGCCCGTCGCT3′ (Harmey & Harmey, 1993) in a PCR reaction. High temperature stable TAQ polymerase (Maniatis *et al.*, 1989) was used in a thermocycle programme of 94°C for 1 min followed by 30 cycles of 94°C for 1 min, 55°C for 1 min and 72°C for 2 min, and one cycle at 72°C for 10 min. The PCR product was ethanol precipitated and sequenced directly using standard procedures. Sequencing was done on a Beckman CEQ 2000 automated sequencer using laser excitation of fluorescent probes. Sequences were compared by computer analysis.

For other experiments the PCR product was separated on a 1.8% agarose gel and the 500 bp band representing the ITS region isolated from the gel using GeneClean™ as described by the manufacturer. The isolated DNA was ligated into PCRII plasmid using the TA cloning kit™ from Invitrogen. The plasmid was then transformed into *E. coli* D10 α. Transformed cells were isolated and amplified. Cloned ITS was removed from the cells by EcoRI restriction and sequenced.

Table 1. *Isolates of* Bursaphelenchus *selected for molecular taxonomy and characterisation.* Bx *indicates* B. xylophilus, R *designate the R and M forms as described by Webster* et al. *(1990),* Bm *indicates* B. mucronatus, *and* av *indicates avirulent isolates. Host species are* Pinus, Abies *and* Larix. *Isolates from the United States are designated by US and those from Canada by C. Isolates from Japan are identified both by city and prefecture of isolation where possible. * Isolate names given by Webster* et al. *(1990).*

Origin	Isolate	Location	Species	Host
North America				
	US-1	Millestone, NJ	BxR	*Pinus thunbergii*
	US-2	Burlington, VT	BxR	*P. strobus*
	US-8	Lee County, IA	BxR	*P. sylvestris*
	US-9	Tucson, AZ	BxR	*P. halepensis*
	US-10	Cloquet, MN	BxM	*Abies balsamae*
	US-11	Burlington, VT	BxR	*Larix larcina*
	US-12	Columbia, MO	BxR	*P. sylvestris*
	US-12B	Eureka, MO	BxR	*P. resinosa*
	US-13	Black River Falls, MN	BxR	*P. resinosa*
	US-14	Georgia	BxM	unknown pine source
	US-15	Maine	BxR	*P. resinosa*
	C-1 (St William)*	Ontario	BxR	*P. sylvestris*
	C-2 (Q1426)*	Quebec	BxM	*A. balsamae*
	C-3 (BC)*	British Columbia	BxR	wood chips
	C-4 (Alberta)*	Alberta	BxR	unknown
	C-5 (mm)*	British Columbia	BxR	*Monochamus* spp.
	C-6 (Q52A)*	Quebec	BxR	wood chips
	C-7 (St John)*	New Brunswick	BxR	wood chips
Europe				
	F-1	Saint Symphorien, France	BxM	*Pinus pinaster*
	F-2	France	Bm	unknown
	D-1	Germany	Bm	unknown
	D-2	Prati di Pra, Germany	Bm	unknown
	N-1	Norway	Bm	unknown
	Fi-1	Finland	Bm	wood chips
	Fi-2	Finland	Bm	wood chips
	Fi-3	Finland	Bm	wood chips
	Sw-1	Sweden	Bm	unknown

Table 1. *(Continued).*

Origin	Isolate	Location	Species	Host
Japan				
	J-2	Nagasaki	BxM	*P. thunbergii*
	J-11	Ichinoseki, Iwate	BxR	*P. densiflora*
	J-12	Tateyama, Chiba	BxR	*Monochamus alternatus*
	J-13	Yachiyo, Chiba	Bm	*P. densiflora*
	J-14	Takahagi, Ibaraki	Bm	*P. densiflora*
	S-10	Hirose, Shimane	BxR	*P. densiflora*
	C-14-5	Ichinomiya, Chiba	BxR av	*P. densiflora*
	OK-2	Onna, Okinawa	BxR av	*M. alternatus*
	S-6-1	Ibaraki	BxR	*P. densiflora*
	S-1	not known	Bm	not known
	T-4	Iwate	BxR	*P. densiflora*

Primers as described by Beckenbach *et al.* (1992, 1999) were used to determine sequence differences in the *hsp-70* gene. PCR conditions were as described above.

Sequences were aligned with the ClustalPPC program for analysis of sequence homologies. Parsimony analysis of sequence data sets was done with PAUP version 3.1.1. This analysis resulted in 24 equal most parsimonious reconstructions of the data. Strict and majority rule trees were developed using *Caenorhabditis elegans* as an out-group.

PATHOGENICITY

Pathogenicity was determined by inoculation of 100 nematodes into a branch on a 6-year-old, field-raised pine and observing wilting symptoms. A population was considered virulent if total wilting occurred, moderately virulent if wilting was confined to the infection site, and resistant if no obvious wilt symptoms occurred.

Results

MATING

The results of single pair matings to produce an F1 generation are shown in Table 2. Although the French isolate (F-1) mated with the

Table 2. *Potential for matings between different isolates of pine wood nematode. Single pair matings were established as described in Materials and methods. + indicates that an F1 generation was established, – indicates no F1 generation was established.*

	US-12	US-10	SW	J-13	J-14	F-1
US-1	+	–	+	+	–	–
US-2	+	–	–	+	–	+
US-8	+	+	+	–	–	+
US-9	+	+	+	–	–	–
US-10	–	+	–	–	–	–
US-11	+	–	–	–	–	+
US-13	+	–	–	–	–	–
US-12	+	+	+	+	–	–
C-1	+	–	–	+	+	–
C-2	+	–	+	–	–	–
F-1	–	–	+	–	–	+
J-2	+	–	+	–	–	–
J-11	+	–	+	+	+	–
J-12	+	+	–	–	–	–
J-13	–	–	–	+	–	–
J-14	–	–	–	–	+	–
OK-2	+	–	+	+	+	–
C-14-5	+	–	–	+	+	–
S-10	+	+	+	–	–	–
B. seani	–	–	–	–	–	–

isolates from Finland (Fi) to produce an F1 generation, there was no F2 generation (Table 3) nor could the F1 progeny of this mating be backcrossed to the parental F-1 or Fi populations (data not shown). There was no clear pattern of mating restrictions between several populations of *B. xylophilus* M or R forms or between those populations described as *B. mucronatus* and those described as *B. xylophilus* within the isolates tested in producing an F-1 generation (Table 2). For example, the population isolated from Sweden reproduced with *B. xylophilus* as well as *B. mucronatus* to produce an F1 generation. The population US-1, a *B. mucronatus* R form population, mated only with other R form populations. As a final example, the population C2 from Canada, described as an M form *B. xylophilus* but clearly a *B. mucronatus* population, mates only with US-12 and R form *B. xylophilus*.

Table 3. *Success of establishment of a population of pine wood nematode derived from an F1 generation established by single pair matings. The potential of establishment of a viable population was based on the mating of F1 males and females. The F1 generations were established by single pair matings as described in Materials and methods. The potential for establishment of a continuing population was considered positive only if the population reproduced by inbreeding for seven generations.*

Original mating	Successful establishment of continuing population
US-1 x US-12	YES
US-1 x C-2	NO
US-1 x C-1	YES
US-1 x J-11	YES
US-1 x J-14	NO
US-2 x J-11	NO
US-2 x J-14	NO
US-14 x US-12	NO
US-14 x J-14	YES
US-12 x J-14	YES
US (ANY) x F-1	NO
J (ANY) x F-1	NO
US (ANY) x Sw	NO
J (ANY) x Sw	NO
C-1 x Sw	NO
C-2 x Sw	YES
C-1 OR C-2 x F-1	NO
Sw x F-1	NO
Fi x F-1	NO

Mating isolation becomes apparent when mating success is determined based on the success of an F2 and subsequent generations or back crossing of the F1 generations to parental generations (Table 3). Successful matings occur only in matings which are *B. xylophilus* R form × *B. xylophilus* R form, *B. xylophilus* M form × *B. xylophilus* M form or *B. mucronatus*, or *B. mucronatus* × *B. mucronatus*. No other F1 generations derived from cross species matings survived to produce an F2 and subsequent generations (Table 3). Backcrosses of the F1s from unsuccessful F2 matings likewise were unsuccessful (data not shown). Most interestingly, no F1 generations derived from matings with the Swedish population could inbreed to produce an F2 generation (Table 3), suggesting a uniqueness of this population.

Chromosome number

Populations clearly separate into three classes based on chromosome number (Table 4). The avirulent Japanese isolates C14-5 and OK-2 along with several *B. xylophilus* isolates of varied virulence in Japanese red, Japanese black or Scots pine, fall into a class with $n = 6$. Several well-identified *B. mucronatus* populations fall into a class with $n = 5$; however this is not indicative of *B. mucronatus* as representative populations fell into classes of $n = 3$ and 6 as well. Additionally no regional pattern of chromosome number was seen (Table 4). When mating behaviour in the lab is analysed as a function of chromosome number (Table 2) the pattern is that no clearer picture is seen if only the F1 generation is considered. If, however, the production of F2 and later generations is considered (Table 3) only those matings between populations from the same chromosome number class are successful as mating pairs.

Virulence

Scots pine and Japanese red pine were most susceptible to PWN infection, being resistant to only *B. mucronatus* populations from Finland and Japan and population F-1 from France. Japanese black pine was highly susceptible to populations of *B. xylophilus* from Japan and moderately susceptible to *B. xylophilus* populations from North America, a *B. mucronatus* population from Germany and the French population F-1. The *B. mucronatus* populations J-13 and J-14 showed no pathogenicity on any pine tested under the testing conditions employed.

Molecular taxonomy

Because of difficulty differentiating *B. xylophilus* from *B. mucronatus* and in describing other *Bursaphelenchus* spp. found in association with dead or dying pines based solely on morphological characters, we turned to differences in gene sequence. The data presented has been derived from work in our laboratory combined by data analysis with published sequences and sequences available in GenBank.

Two separations of populations of PWN become clear when ITS-1 sequence differences are used as a defining character and one of the

Table 4. *Haploid chromosome number of different isolates of* Bursaphelenchus *spp. The N chromosome number was determined as described in Materials and methods as a potential criterion for differentiating isolates and populations within the pine wood nematode species complex.* BxR *indicates the R form isolate of* B. xylophilus, BxM *indicates the M form isolate of* B. xylophilus; BM *refers to* B. mucronatus; *isolates indicated as* av *are considered to be avirulent.*

Isolate	Chromosome number	Isolate identification
US-1	3	BxR
US-9	3	BxR
US-10	3	BxM
US-11	3	BxR
US-13	3	BxR
F-1	3	BM
J-2	3	BM
J-11	3	BxR
J-12	3	BxR
C-2	5	BxM
J-13	5	BM
J-14	5	BM
Fi-1	5	BM
Fi-2	5	BM
US-14	5	BxM
OK-2	6	BxRav
C-14-5	6	BxRav
S-10	6	BxR
C-1	6	BxR
US-2	6	BxR
US-8	6	BxR
US-12	6	BxR
D-1	6	BM
D-2	6	BM
B. abruptus	14	

most parsimonious reconstructions of the data is considered (Fig. 4). The group characterised by US-2, US-12, S6-1, S-10, and OK-2 represents only populations characterised as *B. xylophilus*. The other

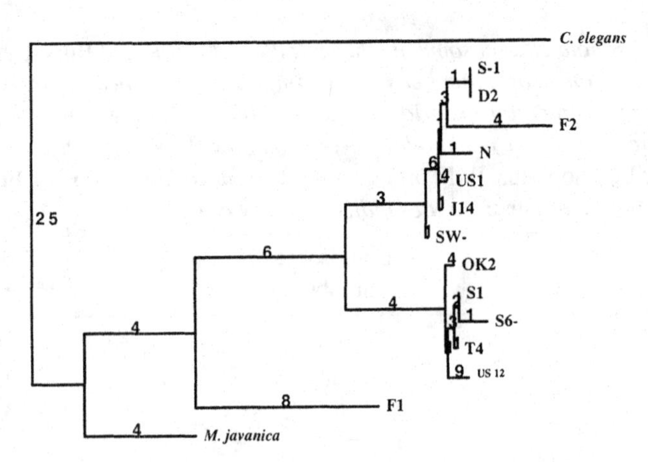

Fig. 4. *Cladogram showing the most parsimonious representation of the relationships between different populations of members of the pine wood nematode species complex based on sequence differences in the ITS I region of ribosomal DNA.* Caenorhabditis elegans *and* Meloidogyne javanica *(root-knot nematode) were used as outlying species. Population designation as in Table 1.*

group containing SW-1, S-1, D-2, F-2, N-1, C-2 and J-14 is a grouping of populations described as *B. mucronatus*. Within this group F-2, a population from France, has a significant number of nucleotide changes from S-1 and C-2 suggesting that it is only distantly related to these other two populations although it may have originally descended from them. In addition, both C-1 from Canada and J-14 from Japan, identified as *B. xylophilus* M form, are within this group, suggesting a close relationship to *B. mucronatus* populations from Sweden, France, Germany and Norway.

Evaluation of taxonomic relationships based on sequence differences in heat shock protein 70 (*hsp 70*) clearly separates a North American/Asian clade from a European clade of *Bursaphelenchus* within the PWNSC (Fig. 5). The North American/Asian clade contains populations described as *B. mucronatus* (US-14 and C-2) and the European clade contains a Portuguese population identified as *B. xylophilus* (Mota *et al.*, 1999; Sousa *et al.*, 2001). Nine Japanese isolates, identical in *hsp 70* sequence, are in the North American/ Asian clade. The type specimen for *B. mucronatus*, population J-13 from Japan, lies outside both descriptive clades (Fig. 5).

Japanese populations can be separated from US/European populations based on 18S rDNA sequence (Fig. 6). This analysis places the

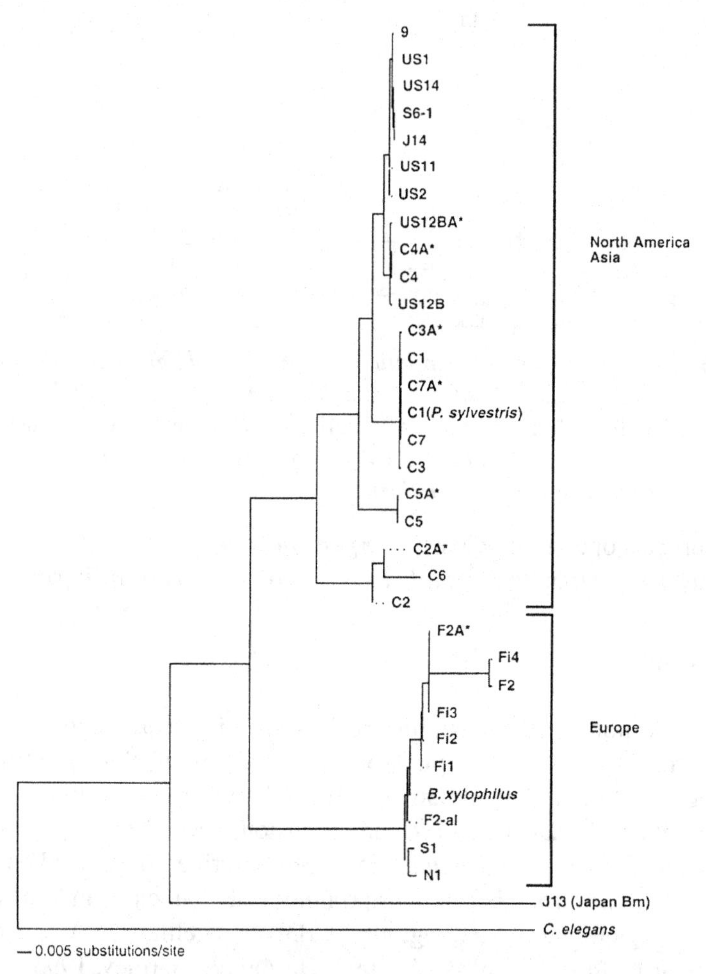

Fig. 5. *Cladogram representing species and population relationships among several members of the pine wood nematode species complex based on sequence differences in the gene sequence encoding the 70S heat shock protein (*Hsp-70*). Identification as in Table 1 with exceptions that (9) indicates a grouping of nine isolates of* Bursaphelenchus xylophilus *from throughout Japan that showed on sequence differences and* B. xylophilus *indicates a population of this nematode from Canada whose origin was not identified in the sequence posting on NIH Blast. Populations noted with * represent comparisons of sequence analysis done in our laboratory to that of Beckenbach* et al. *(1999). No significant sequence difference was detected between these two separate analyses.* Caenorhabditis elegans *was sequenced as an outlying group. The scale for substitution is indicated on the figure.*

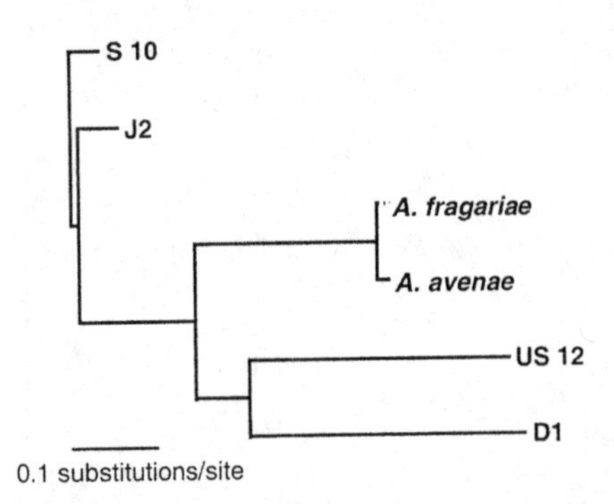

Fig. 6. *Cladogram indicating relationships of four different populations of the pine wood nematode species complex, identified as in Table 1, with* Bursaphelenchus, Aphelenchoides fragariae *and* Aphelenchus avenae. *The limited amount of data limits the validity of this comparison. The scale of sequence differences is shown on the figure.*

US/Europe populations closer to *Aphelenchoides fragariae* and *Aphelenchus avenae* than to *Bursaphelenchus* populations from Japan.

Discussion

The taxonomic status within the PWNSC is clearly confused and points to a complex of populations and species of *Bursaphelenchus* (Steiner & Buhrer, 1934; Baujard, 1980, Webster *et al.*, 1990, Beckenbach *et al.*, 1992). To add to the confusion, several new species and populations of *Bursaphelenchus* have been described from dead and dying conifers in Europe based on morphological characters and on RFLP analysis of internal spacer regions of rDNA, satellite DNA and other molecular methods (Tares *et al.*, 1992; Harmey & Harmey, 1993; Hoyer *et al.*, 1998; Iwahori *et al.*, 1998, 2000; Braasch *et al.*, 1999; Abad, 2000; Bolla & Wood, 2000). Questions to be answered are: *i*) whether these molecular methods used are sufficient to describe species and whether there are really new species associated with dead or dying conifers; *ii*) whether these populations are sibling species of indigenous European species; or *iii*) whether they are species introduced into Europe from North America or Japan.

As one considers the potential for spread of pine wood nematode and pine wilt disease in Europe and Asia, several other questions arise. What are the relationships between populations within the pine wood nematode species complex from North America, Asia and Europe?

What populations, if any, have been introduced into Europe? What is the potential for European populations to interbreed with populations from other areas, and what is the potential then for development of populations pathogenic to European pines? Are populations of *B. xylophilus* presently in Europe?

We have not attempted to answer these questions directly but rather have presented behavioural, cytological and molecular taxonomic approaches that might be used in obtaining answers. Mating potential between isolates offers one criterion to differentiate isolates but only if defined conditions are met. It is clear from the data presented that matings can be forced under laboratory conditions: however, do these matings occur naturally? Riga and Webster (1992) and Bolla *et al.* (1989) demonstrated differential attraction between isolates. This attraction may be pheromone-based and thus may define a biochemical difference between species and perhaps isolates that assure reproductive fidelity. Riga and Webster (1992) demonstrated that population F-1 from France was not attracted to either *B. mucronatus* or *B. xylophilus*. Secondly, it is clear that matings forced in the laboratory do not result in populations that can be sustained. Thus if mating is to be used as a criterion to define isolate relationships, one must consider whether the populations will be brought in contact with each other *via* dispersal by the vector insect, if they will attract for mating to occur in nature, and if the resulting population will sustain itself for several generations. If these criteria are considered in the light of our observations (Bolla *et al.*, 1989) and those of other workers (Mamiya, 1986; Riga *et al.*, 1992) then it is unlikely that any population other than C-2 (Canadian *B. mucronatus*) will mate with any European population. It is equally unlikely that any Japanese populations will mate with European populations. Additionally the French population F-1 is unlikely to mate with either European or North American populations. These results therefore support the concept of a North American/Japanese clade and a European clade of the PWNSC. The results also support F-1 as a separate or sibling species.

Chromosome number can provide an indication of population dynamics but is unlikely, at least alone, to be a reasonable taxonomic tool.

Evolutionary divergence of genome sequence has been widely used as an indication of species relationships and population and evolutionary genetics. Is this then a valid criterion for differentiation within the PWNSC? Several investigators have used a variety of molecular-based approaches to describe pine wood nematode populations and species relationships using RFLP analysis within the rDNA spacer sequences, rDNA and *hsp 70* sequences, and RFLP differences throughout the genome (Tares *et al.*, 1992; Harmey & Harmey, 1993; Hoyer *et al.*, 1998; Iwahori *et al.*, 1998, 2000; Braasch *et al.*, 1999; Abad, 2000; Bolla & Wood, 2000). It is our view that this must be done with caution and data subjected to bioinformatics analysis. The data presented in this paper based on analysis of sequence differences within ITS1 of rDNA, 18S rDNA, and *hsp 70* presents the need for caution as three different interpretations can be derived. More importantly it points to the need for a definition of characters and ecological conditions to be used to define the potential of importation of pine wilt disease into areas where this disease is not present. Careful attention should be paid not only to the population of PWN but to the annual temperature and water profile within the importation region (Rutherford & Webster, 1987; Ikeda *et al.*, 1990; Rutherford *et al.*, 1990, 1992; Kuroda, 1991; Ikeda & Kiyohara, 1995; Ikeda, 1996a, b; Ishida & Hogetsu, 1997; Jikumaru & Togashi, 2000) before decisions to ban importation are made. If these considerations are made the conclusion may be that select rather than global bans on importation of potentially contaminated wood are more appropriate to prevent global spread of pine wilt disease.

In summary, it is clear that no single criterion can stand alone to describe relationships among populations and species within the PWNSC. No single molecular character can be used alone to describe population or species differences; rather, several criteria must be considered. We believe that sequence within a single gene alone is invalid for describing species or populations and that changes in a single base within a sequence cannot be a criterion for differentiation. We recommend that a set of agreed standards, including morphological, biochemical, behavioural and molecular characters, be developed for describing species and populations of *Bursaphelenchus* associated with forest pathology.

References

ABAD, P. (2000). Satellite DNA used as a species-specific probe for identification of *Bursaphelenchus xylophilus*. *EPPO Bulletin* 30, 571-574.

AIKAWA, T. & TOGASHI, K. (2000). Movement of *Bursaphelenchus xylophilus* (Nematoda: Aphelenchoididae) in tracheal system of adult *Monochamus alternatus* (Coleoptera: Cerambycidae). *Nematology* 2, 495-500.

ANBUTSU, H. & TOGASHI, K. (2000). Deterred oviposition response of *Monochamus alternatus* (Coleoptera: Cerambycidae) to oviposition scars occupied by eggs. *Agricultural and Forest Entomology* 2, 217-223.

BAUJARD, P. (1980). Trois nouvelles espèces de *Bursaphelenchus* (Nematoda: Tylenchida) et remarques sur le genre. *Revue de Nématologie* 3, 167-177.

BECKENBACH, K., SMITH, M.J. & WEBSTER, J.M. (1992). Taxonomic affinities and intra- and interspecific variation in *Bursaphelenchus* spp. as determined by polymerase chain reaction. *Journal of Nematology* 24, 140-147.

BECKENBACH, K., BLAXTER, M. & WEBSTER, J.M. (1999). Phylogeny of *Bursaphelenchus* species derived from analysis of ribosomal internal transcribed spacer DNA sequences. *Nematology* 1, 539-548.

BEDKER, P.J., WINGFIELD, M.J. & BLANCHETT R.A. (1987). Pathogenicity of *Bursaphelenchus xylophilus* on three species of pine. *Canadian Journal of Forest Research* 7, 51-57.

BOLLA, J., BRAMBLE J. & BOLLA, R.I. (1989). Attraction of *Bursaphelenchus xylophilus* pathotype MPSy-1 to *Monochamus carolinensis* larvae. *Japanese Journal of Nematology* 19, 32-37.

BOLLA, R.I. & BOSCHERT, M. (1993). Pinewood nematode species complex: interbreeding potential and chromosome number. *Journal of Nematology* 25, 227-238.

BOLLA, R.I. & WOOD, R. (2000). Pathogenicity and potential for spread of pinewood nematode. In: Futai, K., Togashi, K. & Ikeda, T. (Eds). *Sustainability of pine forests in relation to pine wilt and decline. Proceedings of the symposium, Tokyo, Japan, 26-30 October 1998*. Kyoto, Japan, Shokado Shoten, pp. 3-13.

BOLLA, R.I., WINTER, R.E.K., FITZSIMMONS, K. & LINIT, M. J. (1986). Pathotypes of the pinewood nematode *Bursaphelenchus xylophilus*. *Journal of Nematology* 18, 230-238.

BRAASCH, H. (1998). *Bursaphelenchus hofmanni* sp. n. (Nematoda, Aphelenchoididae) from spruce wood in Germany. *Nematologica* 44, 615-621.

BRAASCH, H. (2001). *Bursaphelenchus* species in conifers in Europe: distribution and morphological relationships. *EPPO Bulletin* 31, 127-142.

BRAASCH, H., SWART, A., TRIBE, G. & BURGERMEISTER, W. (1998). First record of *Bursaphelenchus leoni* in South Africa and comparison with some other *Bursaphelenchus* spp. *EPPO Bulletin* 28, 211-216.

BRAASCH, H., METGE, K. & BURGERMEISTER, W. (1999). *Bursaphelenchus*-Arten (Nematoda, Parasitaphelenchidae) in Nadelgeholzen in Deutschland und ihre ITS-RFLP-Muster. *Nachrichtenblatt des Deutschen Pflanzenschutzdienstes* 51, 312-320.

CAROPPO, S., AMBROGIONI, L., CAVALLI, M. & CONIGLIO, D. (1998). Occurrence of the pine wood nematode, *Bursaphelenchus* spp., and their possible vectors in Italy. *Nematologia Mediterranea* 26, 87-92.

CHOI, Y.E. & MOON, Y.S. (1989). Survey on distribution of pine wood nematode (*Bursaphelenchus xylophilus*) and its pathogenicity to pine trees in Korea. *Korean Journal of Plant Pathololog*y 5, 277-286.

DROPKIN, V.H. & FOUDIN, A.S. (1979). Report on the occurrence of *Bursaphelenchus lignicolus*-induced pine wilt disease in Missouri. *Plant Disease Reporter* 63, 904-905.

DWINELL, L.D. (1985). Relative susceptibilities of five pine species to three populations of pinewood nematodes. *Plant Disease Reporter* 69, 440-442.

DWINELL, L.D. (1987). Role of *Bursaphelenchus xylophilus* in the mortality of sand pine in a seed orchard in central Georgia. *Journal of Nematology* 4, 520.

DWINELL, L.D. & NICKLE, W.R. (1989). An overview of the pine wood nematode ban in North America. *USDA Forest Service General Technical Report* SE 55. 13 pp.

EVANS, H., MES-HARTREE, M. & KUBICEK, Q.B. (1993). Risk of transmission of pinewood nematode, its vectors and pine wilt to EC forests. *Report of the EC/Canada/USA Technical Team.* Unpublished report available from Natural Resources Canada, Canada Forest Service, Ottawa, ON, Canada, 25 pp.

EVANS, H.F., MCNAMARA, D.G., BRAASCH, H., CHADOEUF, J. & MAGNUSSON, C. (1996). Pest risk analysis (PRA) for the territories of the European Union (as PRA area) on *Bursaphelenchus xylophilus* and its vectors in the genus *Monochamus. EPPO Bulletin* 26, 199-249.

FIELDING, N.J. & EVANS, H.F. (1996). The pine wood nematode *Bursaphelenchus xylophilus* (Steiner and Buhrer) Nickle (= *B. lignicolus* Mamiya and Kiyohara): an assessment of the current position. *Forestry* 69, 35-46.

HARMEY, J.H. & HARMEY, M.A. (1993). Detection and identification of *Bursaphelenchus* species with DNA fingerprinting and polymerase chain reaction. *Journal of Nematology* 25, 406-415.

HOYER, U. VON, BURGERMEISTER, W. & BRAASCH, H. (1998). Identification of *Bursaphelenchus* species (Nematoda, Aphelenchoididae) on the basis of amplified ribosomal DNA (ITS-RFLP). *Nachrichtenblatt des Deutschen Pflanzenschutzdienstes* 50, 273-277.

IKEDA, T. (1996a). Xylem dysfunction in *Bursaphelenchus xylophilus*-infected *Pinus thungergii* in relation to xylem cavitation and water status. *Annals of the Phytopathological Society of Japan* 62, 554-558.

IKEDA, T. (1996b). Responses of water-stressed *Pinus thunbergii* to inoculation with avirulent pine wood nematode (*Bursaphelenchus xylophilus*): water relations and xylem histology. *Journal of Forest Research* 1, 223-226.

IKEDA, T. & KIYOHARA, T. (1995). Water relations, xylem embolism and histological features of *Pinus thunbergii* inoculated with virulent of avirulent pine wood nematode, *Bursaphelenchus xylophilus*. *Journal of Experimental Botany* 46, 441-449.

IKEDA, T., KIYOHARA, T. & KUSUNOKI, M. (1990). Change in water status of *Pinus thunbergii* Parl. inoculated with species of *Bursaphelenchus*. *Journal of Nematology* 22, 132-135.

ISHIDA, K. & HOGETSU, T. (1997). Role of resin canals in the early stages of pine wilt disease caused by the pine wood nematode. *Canadian Journal of Botany* 75, 346-351.

IWAHORI, H., KANZAKI, N. & FUTAI, K. (2000). A simple, polymerase chain reaction-restriction fragment length polymorphism-aided diagnosis method for pine wilt disease. *Forestry Pathology* 30, 157-164.

IWAHORI, H., TSUDA, K., KANZAKI, N., IZUI, K. & FUTAI, K. (1998). PCR-RFLP and sequence analysis of ribosomal DNA of *Bursaphelenchus* nematodes related to pine wilt disease. *Fundamental and Applied Nematology* 21, 655-666.

JIKUMARU, S. & TOGASHI, K. (2000). Temperature effects on the transmission of *Bursaphelenchus xylophilus* (Nemata: Aphelenchoididae) by *Monochamus alternatus* (Coleoptera: Cerambycidae). *Journal of Nematology* 32, 110-116.

KANZAKI, N., TSUDA, K. & FUTAI, K. (2000). Description of *Bursaphelenchus conicaudatus* n. sp. (Nematoda: Aphelenchoididae), isolated from the yellow-spotted longicorn beetle, *Psacothea hilaris* (Coleoptera: Cerambycidae) and fig trees, *Ficus carica*. *Nematology* 2, 165-168.

KIYOHARA, T. & BOLLA, R.I. (1990). Pathogenic variability among isolates of the pinewood nematode *Bursaphelenchus xylophilus*. *Forest Science* 36, 1061-1076.

KONDO, E., FOUDIN, A., LINIT, M., SMITH, M., BOLLA, R., WINTER, R.I.K. & DROPKIN, V.H. (1982). Pine wilt disease: nematological, entomological and biochemical investigations. *Bulletin SB-372, University of Missouri Agriculture Experimental Station, Columbia, MO, USA*, 56 pp.

KULINICH, O.A., KRUGLIC, I.A., EROSHENKO, A.S. & KOLOSOVA, N.V. (1994). Occurrence and distribution of the nematode *Bursaphelenchus mucronatus* in the Russian Far East. *Russian Journal of Nematology* 2, 113-119.

KURODA, K. (1991). Mechanism of cavitation development in the pine wilt disease. *European Journal of Forest Pathology* 21, 82-89.

MAGNUSSON, C. & KULINICH, O.A. (1996). A taxonomic appraisal of the original description, morphology and status of *Bursaphelenchus kolymensis* Korentchenko, 1980 (Aphelenchida: Aphelenchoididae). *Russian Journal of Nematology* 4, 155-161.

MAMIYA, Y. (1984). The pinewood nematode. In: Nickle, W.R. (Ed.) *Plant and insect nematodes*. New York: Marcel Dekker, pp. 589-628.

MAMIYA, Y. (1986). Interspecific hybridization between *Bursaphelenchus xylophilus* and *B. mucronatus* (Aphelenchida: Aphelenchoididae). *Applied Entomological Zoology* 21, 159-163.

MAMIYA, Y. (1988). History of pine wilt disease in Japan. *Journal of Nematology* 20, 219-226.

MAMIYA, Y. & ENDA, N. (1979). *Bursaphelenchus mucronatus* n. sp. (Nematoda: Aphelenchoididae) from pinewood and its biology and pathogenicity to pine trees. *Nematologica* 25, 353-361.

MOTA, M., BRAASCH, H., BRAVO, M.A., PENAS, A.C., BURGMEISTER, W., METGE, K. & SOUSA, E. (1999). First report of *Bursaphelenchus xylophilus* in Portugal and in Europe. *Nematology* 1, 727-734.

PHILIS, J. (1996). An outlook on the association of *Bursaphelenchus leoni* with wilting pines in Cyprus. *Nematologia Mediterranea* 24, 221-225.

PHILIS, J. & BRAASCH, H. (1996). Occurrence of *Bursaphelenchus leoni* (Nematoda, Aphelenchoididae) in Cyprus and its extraction from pinewood. *Nematologia Mediterranea* 24, 119-123.

RIGA, E. & WEBSTER, J.M. (1992). Use of sex pheromones in the taxonomic differentiation of *Bursaphelenchus* spp. (Nematoda) pathogens of pine trees. *Nematologica* 38, 133-145.

RIGA, E., BECKENBACH, K. & WEBSTER, J. (1992). Taxonomic relationships of *Bursaphelenchus xylophilus* and *B. mucronatus* based on interspecific and intraspecific cross-hybridization and DNA analysis. *Fundamental and Applied Nematology* 15, 391-395.

RÜHM, W. (1956). Die nematoden der Ipiden. *Parasitologische Schriftenreihe Jena* 6, 1-435.

RUTHERFORD, T.A. & WEBSTER, J.M. (1987). Distribution of pine wilt disease with respect to temperature in North America, Japan, and Europe. *Canadian Journal of Forest Research* 17, 1050-1059.

RUTHERFORD, T.A., MAMIYA, Y. & WEBSTER, J.M. (1990). Nematode-induced pine wilt disease: factors influencing its occurrence and distribution. *Forest Science* 36, 145-155.

RUTHERFORD, T.A., RIGA, E. & WEBSTER, J.M. (1992). Temperature-mediated behavioral relationships in *Bursaphelenchus xylophilus, B. mucronatus* and their hybrids. *Journal of Nematology* 24, 40-44.

SAMBROOK, J., FRITSCH, E. F. & MANIATIS, T. (1989). *Molecular cloning: a laboratory manual.* 2nd edition. Cold Spring Harbor, NY, USA, Cold Spring Harbor Laboratory Press, 1659 pp.

SCHAUER-BLUME, M. & STURHAN, D. (1989). Vorkommen von kiefernholznematoden (*Bursaphelenchus* spp.) in der Bundesrepublik Deutschland? *Nachrichtenblatt des Deutschen Pflanzenschutzdienstes* 41, 133-136.

SKARMOUTSOS, G. & MICHALOPOULOS-SKARMOUTSOS, H. (2000). Pathogenicity of *Bursaphelenchus sexdentati, Bursaphelenchus leoni* and *Bursaphelenchus hellenicus* on European pine seedlings. *Forest Pathology* 30, 149-156.

SKARMOUTSOS, G., BRAASCH, H. & MICHALOPOULOU, H. (1988). *Bursaphelenchus hellenicus* sp. n. (Nematoda: Aphelenchoididae) from Greek pinewood. *Nematologica* 44, 623-629.

SOUSA, E., BRAVO, M.A., PIRES, J., NAVES, P., PENAS, A.C., BONIFÁCIO, L. & MOTA, M. (2001). *Bursaphelenchus xylophilus* (Nematoda: Aphelenchoididae) associated with *Monochamus galloprovincialis* (Coleoptera: Cerambycidae) in Portugal. *Nematology* 3, 89-91.

STEINER, G. & BUHRER, E.M. (1934). *Aphelenchoides xylophilus* n. sp., a nematode associate with blue-stain and other fungi in timber. *Journal of Agricultural Research* 48, 949-951.

SUZUKI, K. & KIYOHARA, T. (1978). Influence of water stress on development of pine wilting disease caused by *Bursaphelenchus lignicolus. European Journal of Forest Pathology* 8, 97-107.

TARES, S., ABAD, P., BRUGUIER, N. & GURIAN, G. DE (1992). Identification and evidence for relationships among geographical isolates of *Bursaphelenchus* spp. (pinewood nematode) using homologous DNA probes. *Heredity* 68, 157-164.

WEBSTER, J.M., ANDERSON, R.V., BAILLIE, D.L., BECKENBACH, K., CURRAN, J. & RUTHERFORD, T.A. (1990). DNA probes for differentiating isolates of the pinewood nematode species complex. *Revue de Nématologie* 13, 255-263.

WINGFIELD, M.J., BEDKER, P.J. & BLANCHETTE, R.A. (1986). Pathogenicity of *Bursaphelenchus xylophilus* on pines in Minnesota and Wisconsin. *Journal of Nematology* 18, 44-49.

WINGFIELD, M.J., BLANCHETTE, R.A. & KONDO, E. (1983). Comparison of the pinewood nematode, *Bursaphelenchus xylophilus*, from pine and balsam fir. *European Journal of Forest Pathology* 13, 360-372.

YANG, B.J., LIU, W., XU, F.Y. & ZHANG, P. (2000). The potential threat of pine wilt disease to China forest and its early diagnosis. In: Futai, K., Togashi, K. & Ikeda, T. (Eds). *Sustainability of pine forests in relation to pine wilt and decline. Proceedings of the symposium, Tokyo, Japan, 26-30 October 1998*. Kyoto, Japan, Shokado Shoten, pp. 261-265.

YIN, K., FANG, Y. & TARJAN, A.C. (1988). A key to species in the genus *Bursaphelenchus* with a description of *Bursaphelenchus hunanensis* sp. n. (Nematoda: Aphelenchoididae) found in pinewood in Hunan Province, China. *Proceedings of the Helminthological Society Washington* 55, 1-11.

YOSHIMURA, A., KAWASAKI, K., TAKASU, F., TOGASHI, K., FUTAI K. & SHIGESADA, N. (1999). Modeling the spread of pine wilt disease caused by nematodes with pine sawyers as vector. *Ecology* 80, 1691-1702.

Nematology Monographs & Perspectives, 2003, Vol. 1, 55-64

The pine wood nematode: implications of factors past and present for pine wilt disease

John M. WEBSTER

Department of Biological Sciences, Simon Fraser University, Burnaby, Vancouver, Canada

Summary – Pine wilt disease, although considered an economically important problem mostly in Asia, is an ancient disease that evolved probably in the vast forests of the warmer parts of North America (de Guiran *et al.*, 1985) and subsequently spread to other parts of the globe through world trade. Although this biological interaction is not now a disease of consequence in North America, it is the cause of major economic loss in several Asian countries by killing millions of pine trees every year, and by restricting the trading of lumber and wood products in many other parts of the world. Associated with the disease are the indirect economic losses due to necessary restrictive quarantine regulations and to the modification of forestry practices. The aesthetic loss of dying pines despoils home gardens and parks, as well as devastating the hills and valleys of the wilder countryside. The loss of large numbers of trees also causes significant environmental perturbation by locally diminishing the quality of the soil and eroding it. In addition, the removal of such large numbers of trees from a forested area significantly changes species richness in the forest due to a cascade of changes that affect other flora and macro- and microfauna. On a large scale such transformations lead to changes in the local climate and air quality. Overall, these losses are profound sociologically and economically. There is now growing concern in an increasing number of countries that the spread of the disease is unstoppable.

Bursaphelenchus xylophilus, the causative agent of the disease, is indigenous to North America where its vector, *Monochamus* spp., also occurs (Mamiya, 1983). The symptoms of pine wilt disease were probably recorded first in Japan by Yano (Kishi, 1995) in 1905, although the cause of this disease (Kiyohara & Tokushige, 1971) and an understanding of the role of the *Monochamus* vector and the blue-stain fungus was not recognised until much later (see Mamiya, 1988). It

Table 1. *Biological factors and attributes favouring pine wilt disease.*

Factor	Attribute
Nematode pathogens	large reproductive potential life cycle resilient to adverse conditions by having phytophagous and mycophagous cycles
Susceptible pine host	large areas of forest and/or dense plantations monoculture exotic plantings
Insect vector	dispersal over long distance (up to 20 km flight) up to two generations per year
Alternative fungal host	several species as saprophytes on logs and dying trees

Table 2. *Physical factors favouring pine wilt disease.*

Factor	Effect
High temperatures	nematodes reproduce faster more nematodes per vector nematodes disperse faster in the trees insect vector two generations per year insect flies faster and further trees wilt faster
Low soil moisture	trees wilt faster
Fire	additional stress for the trees

is the current rapid spread of this disease in China, its relatively recent occurrence in several other Asian countries and its recent detection on the European continent, in Portugal (Mota *et al.*, 1999), which has caused increased concern. We now recognise that there are many large areas of susceptible tree species worldwide that are vulnerable to pine wilt disease due to the mutual occurrence of *B. xylophilus* and its *Monochamus* vector, together with fungus in warmer climatic regions. A range of biological (Table 1) and physical factors (Table 2) combine to cause the disease, but it is the persistent practice of certain human

Table 3. *Human factors favouring pine wilt disease.*

Factor	Effect
Worldwide trade	nematode infected logs, lumber and wood products vector infested logs, lumber and wood products barked logs and low grade lumber dunnage infected with the nematode and its vector
Planting of exotic susceptible pine plantations	
Low levels of forest management	

factors (Table 3) that is increasing the chance and rate of spread of the disease (Webster, 1999). Some trading practices and the practice of establishing large plantations of exotic *Pinus* species that are susceptible to pine wilt disease are especially of concern. Several countries in the Southern Hemisphere (*e.g.*, Australia, Chile, New Zealand and South Africa), in particular, need to take exceptional precautions to preclude the nematode and its vector (Webster, 1999).

Canada, a land of softwood forests

Canada, where *B. xylophilus* is indigenous, is a country of vast temperate forests, much of which is pine. There are also several other gymnosperm species in Canada and the United States in which *B. xylophilus* occurs, and the products of these species and of pine have been traded for centuries worldwide. Sutherland and Peterson (1999) summarised the results of an extensive survey of the distribution of *B. xylophilus*, its vectors and potential hosts in Canada. Of 3700 trees examined, less than 10% yielded *B. xylophilus*, and it occurred in scattered pockets of concentration across virtually the whole of the country. *Bursaphelenchus xylophilus* was found in several *Pinus* spp., in Balsam fir *Abies balsamea*, some *Picea* species, larch *Larix laricina* and Douglas fir, *Pseudotsuga menziesii*, but not in *Tsuga* or *Thuja* species. Several *Monochamus* spp., the vectors of *B. xylophilus*, occur in Canada, *e.g.*, *M. clamatus*, *M. obtusus*, *M. notatus* and *M. scutellatus*.

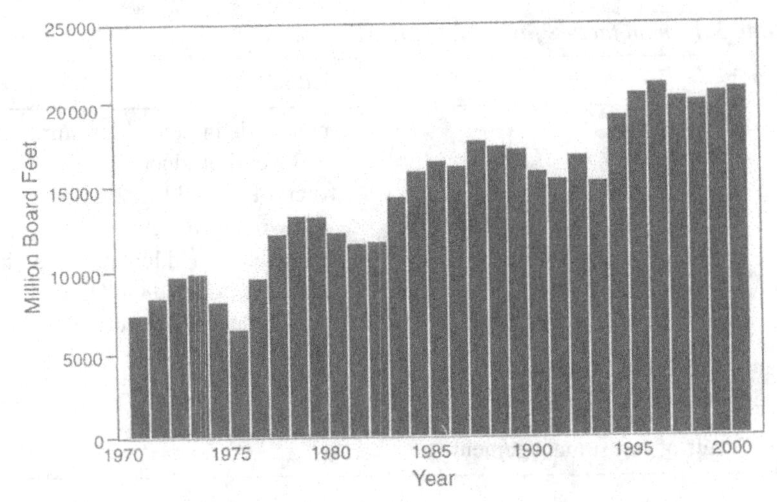

Fig. 1. *Total softwood lumber exports (millions of board feet) from Canada, 1970-2000.*

However, of several thousand *Monochamus* spp. examined in Canada by Sutherland and Peterson (1999), only a small percentage contained *B. xylophilus*. In summary, the biological agents of the disease occur in Canada, but the conditions do not occur for other than isolated localised outbreaks of the disease.

Large quantities of softwood are exported from Canada annually as cut lumber (Fig. 1). Most of this is a mixture of spruce, fir and hemlock and over 80% of it is exported to the United States. Most of the remainder goes to Japan, Korea and China, and only a very small amount now is exported to Europe (Fig. 2). This pattern has changed little over the last 30 years although the overall volume of lumber exports has steadily increased. Every year several hundred thousand m^3 are exported as barked logs (Fig. 3) and, surprisingly, the quantity of log exports have increased significantly in the last few years, especially to Japan and the United States.

Global perspective on trade in softwood

Lumber and wood products are major traded commodities worldwide and originate in the tropical and temperate rainforests. It seems certain that it was the extensive export of softwood from North America that by chance seeded Asia with *B. xylophilus* towards the end of the 19th

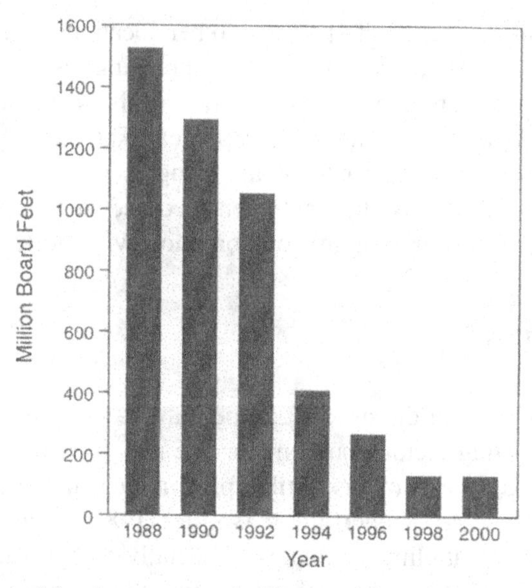

Fig. 2. *Softwood lumber exports (millions of board feet) from Canada to Europe, 1988-2000.*

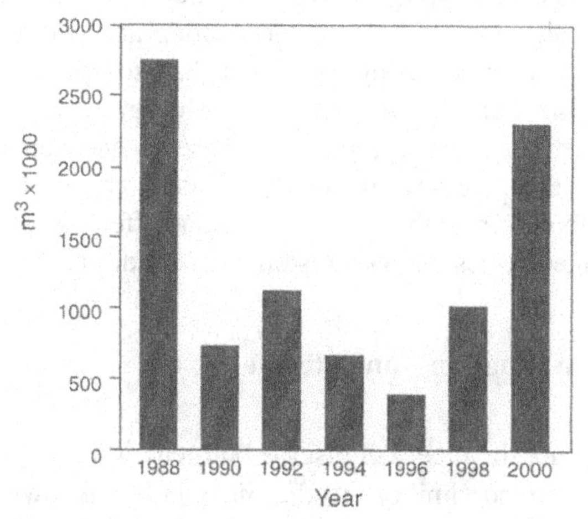

Fig. 3. *Barked log exports (m^3) from Canada, 1988-2000.*

century. This surmise is based on the localised, initial recordings of pine wilt disease in Japan in the first decade of the 20th century, and on the progressive spread of the disease since then throughout most of the higher temperature lowlands of the Japanese archipelago, but excluding the cooler northern island of Hokaido. It was only a matter of time, therefore, before it spread to other locations in Asia through analagous trading practices. It is now spreading rapidly throughout southern China where it has been recorded in five mainland provinces since its first recording in the mid-1980s near Nanjing, adjacent to the great commercial trading route of the Yangtze river. Curiously, until recently, pine wilt disease had not been recorded in Europe despite an even longer period of trading in wood products with North America.

Climatic factors

The absence of the disease in Europe until recently is probably due to many interacting factors, but a major one may have been the relative absence of large forests of susceptible pine in areas that are hot enough to sustain the disease (Rutherford & Webster, 1987). Many pundits and others speculate that climatic change, especially global warming, could result in dire consequences for the temperate forests of Europe as they would more likely be vulnerable to pine wilt disease. Whether or not long-term global warming is occurring is a moot point (Schlaepfer, 1993) as there is localised evidence for both cooling and warming (Table 4). Nevertheless, if it is occurring, pine wilt disease will be one of many factors that take their toll of the forests. Moreover, it is high summer temperatures during the reproductive phase of the nematode and its vector, rather than the severity of the winters, that is more likely to influence the build-up of their populations and affect the stress tolerance of the host pine species, thereby resulting in the disease.

The European situation – an outcome

Whether or not the pine wilt disease outbreak in Portugal will sustain itself so as to spread similarly to that in Asia is unknown at this time. However, if it did, the destruction of what remains of the susceptible *Pinus* species in the hotter parts of Europe within the next half century would be a possible outcome. Fortunately, having been sensitive to the progression of the disease in Asia, scientists and authorities in Europe as a whole, and Portugal in particular, acted rapidly. As well, it seems probable that there are some biotic and abiotic factors in Europe that may tend to localise this outbreak of pine wilt disease.

Table 4. *Recent observations on climate change.*

Global warming?	Global cooling?
1990 the warmest decade in the past 100 years	
1999 global mean combined land surface and sea surface temperature 0.3-0.4°C above 30 year norm	1999 January, coldest in Norway and western Russia in over 100 years
the 21st consecutive year with above normal global surface temperature	February snowfalls were the heaviest in several decades in central and western Europe
the warmest year in England in 341 years	snowfall on Mount Baker (western USA) set a new record (26.96 m)
the 3rd warmest year in Japan in 102 years	greatest one-day snowfall at Tahsta Lake (western Canada) – 145 cm
the warmest year of the century in Russia	
Global mean temperature at end of 20th century 0.7°C higher than at the beginning	

For a parasitic disease to reach epidemic proportions, which could be regarded as the situation of pine wilt disease in Asia, several factors are essential. There must be a minimum density of pines below which populations of the nematode and vector are unlikely to sustain the disease (see Yoshimura *et al.*, 1999) over long periods. The biotic and abiotic conditions must be optimal on an annual, seasonal basis for a sufficiently long period to enable the build-up of high populations of both the nematode and its vector to levels that sustain the disease in a large, contiguous area of susceptible host *Pinus* spp. There are very large areas of susceptible *P. sylvestris* in the cooler regions of northern Europe and there are areas of susceptible pines (*e.g.*, *P. pinaster*) in many parts of the hotter, southern regions of Europe (Critchfield & Little, 1966). Undoubtedly, the hotter (>20°C) climatic conditions that are conducive to pine wilt disease occur in many parts of southern Europe (Rutherford & Webster, 1987). The unanswered question is whether the average

Table 5. *Summary of general factors influencing pine wilt disease by geographic region.*

Organism	Status	Geographic region		
		North America	Asia	Europe
Bursaphelenchus xylophilus	agent	Indigenous	Exotic	Localised exotic
Monochamus spp.	vector	Several indigenous species	Several indigenous species	Several indigenous species
Pinus spp.	host	Large forests of susceptible and resistant species. Few areas of susceptible forests in hotter* localities. Some small areas of exotic pine species	Large forests of susceptible and resistant species in hotter locations	Large forests of susceptible and resistant species. Areas of susceptible tree species mostly in cooler regions but significant scattered forests of susceptible species in hotter areas

* Protracted mean summer air temperature > 20°C

daily temperature remains sufficiently high for long enough over several consecutive summers in large enough areas of susceptible pine forest for the pine wood nematode and its insect vector to sustain populations at a sufficiently high level to reach epidemic rather than acute proportions (Table 5). It is the sustaining of high populations of *B. xylophilus* and its vector over time and the density of forest pines that will determine whether this outbreak of pine wilt disease in Portugal will become an epidemic and spread or remain at acute levels and collapse within a few years. A possible earlier example of a putative, unsustainable population of *B. xylophilus* in Europe is that of the controversial record of this nematode in *P. pinaster* in the Landes forest of south-eastern France (Scotto la Massesse, 1974). This population of *B. xylophilus* appears to have collapsed relatively quickly, and the pine wilt disease did not spread (de Guiran & Bruguier, 1989).

The fact that there is no record of the spread of pine wilt disease in Europe from previous introductions, despite extensive trade over a long period, suggests that the conditions in Europe are sufficiently different from those in Asia for an epidemic not to occur. In addition, the exceptional response of the Portuguese scientific and government authorities and the European legislative bodies at an early stage of occurrence of the disease are helping to mitigate the spread of this economically important disease in Europe.

Acknowledgements

The author thanks Statistics Canada and the Natural Sciences and Engineering Research Council.

References

CRITCHFIELD, W.B. & LITTLE JR, E.L. (1966). Geographic distribution of the pines of the world. *USDA Forest Service Miscellaneous Publications*, No. 991, 97 pp.

DE GUIRAN, G. & BRUGIER, N. (1989). Hybridization and the phylogeny of the pinewood nematode (*Bursaphelenchus* spp.). *Nematologica* 35, 321-330.

DE GUIRAN, G., LEE, M.J., DALMASSO, A. & BONGIOVANNI, M. (1985). Preliminary attempt to differentiate pinewood nematodes (*Bursaphelenchus* spp.) by enzyme electrophoresis. *Revue de Nématologie* 8, 85-92.

KISHI, Y. (1995). *The pinewood nematode and the Japanese pine sawyer.* Tokyo, Japan, Thomas Company Ltd, 302 pp.

KIYOHARA, T. & TOKUSHIGE, Y. (1971). Inoculation experiments of a nematode, *Bursaphelenchus* sp., onto pine trees. *Journal of Japanese Forestry Science* 53, 210-218.

MAMIYA, Y. (1983). Pathology of the pine wilt disease caused by *Bursaphelenchus xylophilus. Annual Review of Phytopathology* 21, 201-220.

MAMIYA, Y. (1988). History of pine wilt disease in Japan. *Journal of Nematology* 20, 219-226.

MOTA, M.M., BRAASCH, H., BRAVO, M.A., PENAS, A.C., BURGERMEISTER, W., METGE, K. & SOUSA, E. (1999). First report of *Bursaphelenchus xylophilus* in Portugal and in Europe. *Nematology* 1, 727-734.

RUTHERFORD, T.A. & WEBSTER, J.M. (1987). Distribution of pine wilt disease with respect to temperature in North America, Japan and Europe. *Canadian Journal of Forest Research* 17, 1050-1059.

SCHLAEPFER, R. (ED.) (1993). *Long-term implications of climate change and air pollution on forest ecosystems.* Progress Report of the International Union of Forestry Research Organisations Task Force "Forest, climate change and air pollution." Vienna, Austria, *IUFRO World Series* 4, 132 pp.

SCOTTO LA MASSÈSSE, C. (1979). Item 3.3. *Report of the joint EPPO/IUFRO Working Party meeting on phytosanitary problems in forestry, Vienna, 1979.* Paris, France, EPPO, Series C, no 60.

SUTHERLAND, J.R. & PETERSON, M.J. (1999). The pinewood nematode in Canada: history, distribution, hosts potential vectors and research. In: Futai, K., Togashi, K. & Ikeda, T. (Eds). *Sustainability of pine forests in relation to pine wilt and decline. Proceedings of the symposium, Tokyo, Japan, 26-30 October 1998.* Kyoto, Japan, Shokado Shoten, pp. 247-253.

YOSHIMURA, A., KAWASAKI, K., TAKASU, F., TOGASHI, K., FUTAI, K. & SHIGESADA, N. (1999). Modeling the spread of pine wilt disease caused by nematodes with pine sawyers as vectors. *Ecology* 80, 1691-1702.

WEBSTER, J.M. (1999). Pine wilt disease: a worldwide survey. In: Futai, K., Togashi, K. & Ikeda, T. (Eds). *Sustainability of pine forests in relation to pine wilt and decline. Proceedings of the symposium, Tokyo, Japan, 26-30 October 1998.* Kyoto, Japan, Shokado Shoten, pp. 254-260.

Nematology Monographs & Perspectives, 2003, Vol. 1, 65-75

Surveys for the pine wood nematode in Russia

Oleg KULINICH

All-Russian Research Institute of Plant Quarantine, Pogranichnaya Str. 32, Moscow Region, 14050 Russia

Summary – The present paper is a short review of research on the pine wood nematode in Russia. Beginning in the late 1980s, coniferous forests in different regions of European and Asian Russia were surveyed for pine wood nematode infections. Only populations of the nematode *Bursaphelenchus mucronatus* were detected. Thirteen populations of *B. xylophilus*, *B. mucronatus, B. kolymensis* and *B. fraudulentus* from Canada, France, Japan, Norway, Sweden, Germany, Russia and the United States were studied for morphological comparisons. Drawings of female tails, bursa, spicules and mucro are shown. Cross-hybridisation among Russian and French isolates of *B. mucronatus* and the Canadian isolate of *B. xylophilus* were conducted and hybrids were produced. Several pathogenicity tests using different isolates of *B. mucronatus* have been conducted in the Russian Far East and European part of Russia. *Bursaphelenchus mucronatus* was pathogenic to conifer species in some tests.

This presentation is a short review of research on the pine wood nematode in Russia. This information is based on research of the pine wood nematode that has been conducted by Russian scientists and does not include the data of foreign researchers who have worked with nematode isolates from Russia.

Distribution of *Bursaphelenchus* species

Studies on pine wood nematode in Russia began in the late 1980s. Diseased pines (*Pinus* sp.) in some central regions of European Russia were surveyed for nematodes with emphasis on species of the genus *Bursaphelenchus.* Extracted nematodes were identified. We also attempted to culture nematodes that were extracted in greatest abundance from the wood samples. The purpose of this work was to understand the role of nematode species in diseases of pine trees. The nematode *Sych-*

notylenchus sp. has been extracted from the wood of Scots pine (*Pinus sylvestris*). The abundance of this species was 200 nematodes per g of wood. The trees on our forested research site were diseased and all wood samples collected contained *Sychnotylenchus* sp. (Kulinich, 1991). It has been proposed that other nematode species may be as pathogenic to pines as some *Bursaphelenchus* species. Unfortunately, our attempts to culture the extracted nematode (*Sychnotylenchus* sp.) have not been successful.

Bursaphelenchus xylophilus is distributed throughout Japan and eastern China and could be transmitted by its beetle vector from China to the Primorsky Krai, Chita or Khabarovsk regions of Russia. Since 1991, the conifer forests of Primorsky Krai have been surveyed to determine if populations of *B. xylophilus* were present. Wood samples from dead and dying spruce, larch, pine and fir were analysed. The survey results indicated that pine wood nematode (*B. xylophilus*) was not present in this region, but populations of *B. mucronatus* were widespread throughout the conifer forests of Primorsky Krai (Kulinich *et al.*, 1994). In 1995-1997, coniferous forests in the European parts of Russia and in the Ural Mountains were also surveyed for pine wood nematode infections. Only populations of the nematode *B. mucronatus* were detected within the following regions: Komy Republic, Sverdlovsk District, Nijni-Novgorod District and the Voronezh District (Kulinich & Orlinskii, 1998a).

According to recent information (Braasch *et al.*, 2001), several *Bursaphelenchus* species were found in wood imported from Russia. Also the pine wood nematode (*B. xylophilus*) was extracted from larch wood from the Krasnoyarsk region of Siberia. However, more data, including DNA analyses, are needed to confirm this identification.

Taxonomic and morphological studies

Significant problems associated with pine wood nematode identification include the taxonomic and current diagnostic techniques used to differentiate this nematode from related species. Morphologically similar species within the genus *Bursaphelenchus* (*B. xylophilus*, *B. mucronatus, B. kolymensis* and *B. fraudulentus*) have several characters that overlap. Thirteen populations of the species previously mentioned were obtained from various countries (Canada, France, Japan, Norway, Sweden, Germany, Russia and the United States) for morphological comparisons conducted in Sweden and Russian laboratories (Magnusson & Kulinich,

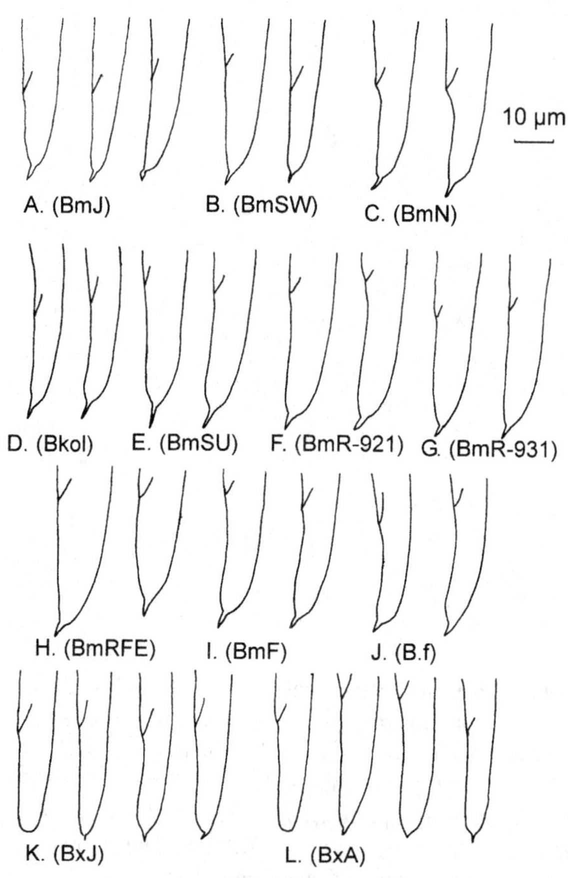

Fig. 1. *Morphology of female tails of isolates of* Bursaphelenchus xylophilus, B. mucronatus, B. kolymensis *and* B. fraudulentus. *A:* B. mucronatus, *Japan (*BmJ*); B:* B. mucronatus, *Sweden (*BmSW*); C:* B. mucronatus, *Norway (*BmN*); D:* B. kolymensis *Russia (*Bkol*); E-H:* B. mucronatus, *Russia (*BmSU, BmR-921, BmR-931, BmRFE*); I:* B. mucronatus, *France (*BmF*); J:* B. fraudulentus, *Germany (*B.f*); K:* B. xylophilus, *Japan (*BxJ*); L:* B. xylophilus, *USA (*BxA*).*

1996). The drawings of female tails, bursa, spicules and mucro for isolates of *B. xylophilus* from USA (BxA) and Japan (BxJ); *B. mucronatus* from France (BmF), Japan (BmJ, BmJ-M), Norway (BmN), Sweden (BmSW), and Russia (BmSU, BmR-1, BmR- 921, BmR-931, BmRFE); *B. kolymensis* (Bkol) and *B. fraudulentus* (B.f) from Germany are shown in Figs 1-3. The Russian *B. mucronatus* isolates were from Siberia and the Russian Far East. A type specimen of *B. kolymensis* was studied.

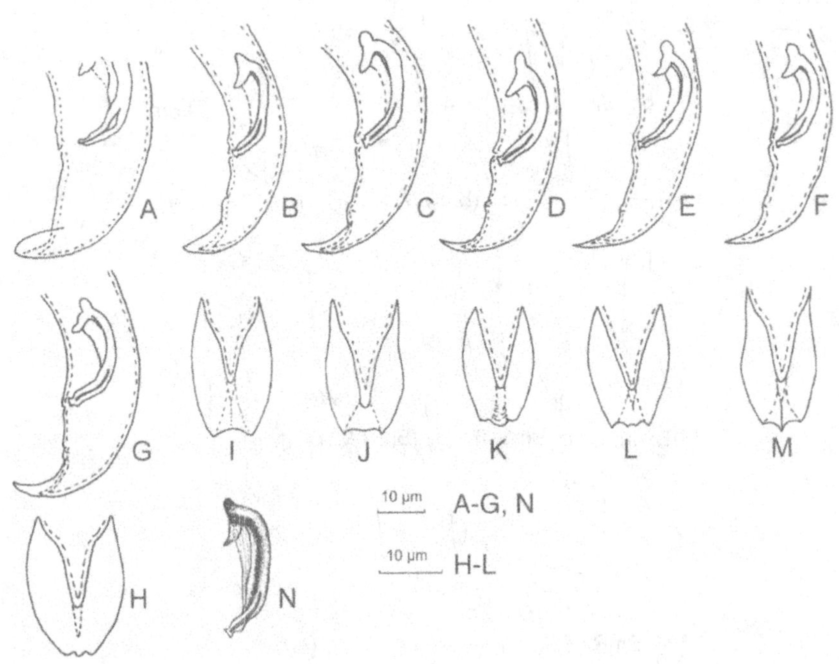

Fig. 2. *The bursa form for* Bursaphelenchus xylophilus, B. mucronatus, B. kolymensis, B. fraudulentus *and their spicules. A-G: In lateral view; I-M: In dorso-ventral view; A:* B. kolymensis, *Russia (*Bkol*); B, I:* B. mucronatus, *Russia (*BmSU*); C, G:* B. mucronatus, *France (*BmF*); D, K:* B. mucronatus, *Japan (*BmJ*), E, L:* B. xylophilus, *Japan (*BxJ*); F, M:* B. xylophilus, *USA (*BxA*), G, H:* B. fraudulentus, *Germany (*B.f*); N: The common spicule for all males in the pine wood nematode species complex.*

Eighteen morphological and allometric parameters have been analysed. Body length, spicule length, stylet length, the indexes c and c′ ($P < 0.05$), bursal shape, female tail shape and the shape of the mucro showed the lowest variability (Kulinich *et al.*, 1996).

Observations on the morphology of the bursa in dorso-ventral view showed that its shape in lateral view depends on the development of the axial rostrum and the bursal alae. According to the development of these parameters (axial rostrum and the bursal alae), three bursa shapes were distinguished in lateral view: beak-shaped, broadly shaped and spade-shaped.

The spicules were removed from the male body, then separated one from the other and studied. Spicule lengths were 17-35.7 µm and were dependent on the body length of the male. All males of the species

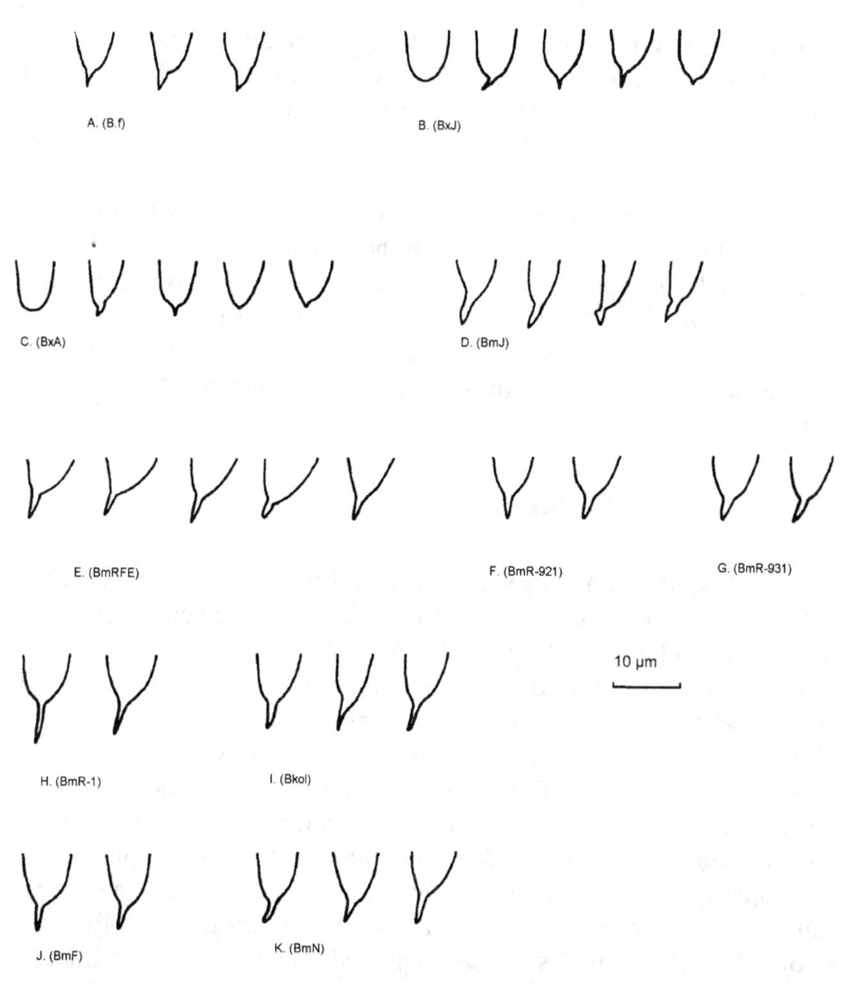

Fig. 3. *Morphology of female mucro of* Bursaphelenchus xylophilus, B. mucronatus, B. kolymensis *and* B. fraudulentus. *A:* B. fraudulentus, *Germany (*B.f*); B:* B. xylophilus, *Japan (*BxJ*); C:* B. xylophilus, *USA (*BxA*); D:* B. mucronatus, *Japan (*BmJ*); E-H:* B. mucronatus, *Russia (*BmRFE; BmR-921, BmR-931, BmR-1*); I:* B. kolymensis *Russia (*Bkol*); J:* B. mucronatus, *France (*BmF*); K:* B. mucronatus, *Norway (*BmN*).*

B. xylophilus, B. mucromatus, B. fraudulentus and B. kolymensis had the same shape spicules (Fig. 2).

The projection of the spicules under the microscope is dependent on body orientation, the focal plane, and the quality of the specimen. Three spicule shapes were distinguished among studied nematodes:

conic, subcylindrical and cylindrical. Also several mucro tail shapes have been distinguished on females. Monofactorial, polyfactorial and cluster analysis of the morphological data demonstrated similarity among Russian isolates of *B. mucronatus* and their difference from *B. mucronatus* of Japan.

Studies of *B. kolymensis*, described from the Maragan district of Russia, indicate that this nematode is morphologically similar to *B. fraudulentus* and *B. mucronatus*, but distinct from *B. xylophilus*. *Bursaphelenchus kolymensis* is most similar to *B. mucronatus* populations found in Russia and France, but differs from Japanese populations because of its female tail, slender mucro and shape of the bursa (Magnusson & Kulinich, 1996).

Results of cross-hybridisation

Cross-hybridisation between Russian and French isolates of *B. mucronatus* and the Canadian isolate of *B. xylophilus* were conducted in the laboratory (Kulinich & Orlinskii, 1998a; Kulinich *et al.*, 1998). The following isolates were used for cross-hybridisation: Russian isolates of *B. mucronatus*, BmKOMY (Komy Republic, *Picea*), French isolates of *B. mucronatus*, BmF (France, *Pinus pinaster*), and Canadian M-form isolates of *B. xylophilus* BxQ-52A M (Quebec, *Abies*). The purpose of these studies was to determine if hybridisation could occur between Russian isolates of *B. mucronatus* and other isolates of the pine wood nematode species complex from different regions of the world and to estimate the potential threat of these hybrids to conifers grown in Russia.

Based on these studies it was suggested that isolates from Canada (BxQ-52A M) and Europe (France and Komy Republic of Russia) are separate species. The females of all isolates of *B. mucronatus* and *B. xylophilus* did not reproduce by parthenogenesis. The experiments showed that hybridisation between populations of *B. mucronatus* from Russia and of *B. xylophilus* (M-form isolate) from Canada is possible.

Pathogenicity

The pathogenicity of Russian *B. mucronatus* isolates has been tested by various researchers. *Bursaphelenchus mucronatus* is considered a non-pathogenic or slightly pathogenic species in pines. However, some research studies conducted using French, Norway and German isolates suggest this species of nematode may be pathogenic in conifers (Panesar & Sutherland, 1989; Braasch, 1994; Mamiya, 1999).

Several pathogenicity tests have also been conducted by Russian researchers. Based on surveys of the distribution of *B. mucronatus* in the Russian Far East (Primorski Krai), it was proposed that this species of nematode was associated with dead *P. sylvestris* and *P. koraiensis* (Kulinich *et al.*, 1994). Further experiments (Eroshenko & Kruglik, 1996) with these isolates conducted in the Russian Far East indicated a strong pathogenic effect on two local conifers, *P. koraiensis* and *Larix olgen.* A less pronounced effect was observed in *P. sylvestris* and *P. densiflora* exposed to these isolates. Morphologically, the Russian Far East isolate is similar to the European isolates, particularly the French isolate.

The level of pathogenicity varies depending on climatic and various biological conditions. The biological cycle of *B. mucronatus* includes a parasitic stage of the nematode in healthy conifers. Although *B. mucronatus* has had a pathogenic effect on some coniferous species in the laboratory, it does not cause extensive damage to native conifers in Europe and Asia. This research may suggest that *B. mucronatus* in Europe and Asia exhibits traits that are similar to those exhibited by *B. xylophilus* species in North America. *Bursaphelenchus xylophilus* does not damage local coniferous species in North America.

Two field experiments were conducted to determine the pathogenicity and survivability of various isolates of *B. mucronatus*, M-form isolates *B. xylophilus*, and their hybrids on *P. sylvestris* in the European part of Russia. Two-year-old seedlings growing in the Moscow region were inoculated in July 1996 with one of two *B. mucronatus* isolates from Russia and one from China (Kulinich *et al.*, 1999). The nematodes were extracted 2 months after seedling inoculation and again 1 year post-inoculation. No nematodes were found in seedling roots or soil. No seedlings had died 14 months after inoculation and only nematodes of the Russian Far East isolate were extracted from seedling stem samples (11 individuals/g of stem tissue).

In another field experiment (1997-1998) in the Voronej region of Russia, 5-year-old *P. sylvestris* seedlings were inoculated with M-form isolates of *B. xylophilus* from Canada, *B. mucronatus* from Russia and France and their hybrids. Nearly all of these isolates survived natural climatic conditions 1 year after inoculation, but no seedlings died (Kulinich & Orlinskii, 2001).

Climate

Climatic conditions conducive to pine wilt disease, caused by the nematode *B. xylophilus*, are associated with sites where the mean temperature for July exceeds 20°C. The warmest summer month in Russia is July. Pine wilt disease would occur in sites found south of this isotherm. There are large areas within Russia where the mean July temperature exceeds 20°C. These sites would include a substantial area of European Russia, from the Black Sea to the Voronez district, and western portions of the Russian Far East, including the Primorsky territory and Khabarovsk district (Kulinich & Kolossova, 1995; Kulinich & Orlinskii, 1998b).

Host plants

The pine wood nematode is pathogenic to several conifer species native to Europe and Asia. *Pinus sylvestris* is the most susceptible pine for pine wood nematode infection. It is widely distributed in Europe and also can be found in all Russian regions. The southern border of *P. sylvestris* distribution extends into the 20°C isotherm throughout numerous regions in the European and Asian sections of Russia. Principal native conifers include: *P. sylvestris, P. sibirica, P. koraiensis, P. pumila, Picea exelsa, P. obovata, P. ajanensis, P. koraiensis, Abies sibirica, A. nephrolepis, A. holophylla, A. sachalinensis, Larix sibirica,* and *L. daurica*. These tree species comprise a major portion of Russia's coniferous forests and are established across large landscapes. The majority of these species have a southern boundary of distribution which does not cross the 20°C July isotherm (Kulinich & Kolossova, 1995). Over 40 tree species have been introduced from North and Central America, Central Asia and the Mediterranean including: *Pinus strobus, P. banksiana, P. pinea, P. halepensis, P. montezumae, P. coulteri, P. sabiniana* and others.

Pine wood nematode vectors

The primary vectors of *B. xylophilus* and *B. mucronatus* are the cerambycid beetles of the genus *Monochamus*. Also other coleopterans (Buprestidae and Curculionidae) have been found to carry nematodes in or on their bodies. Of the 25 species of potential beetle vectors (Linit, 1988, 1990), 11 species are found in Russia. The majority of cerambycid beetles and both species of Curculionidae, although widely distributed throughout North America, are not widespread in Russia (Kulinich & Orlinskii, 1998b).

The primary vector of the pine wood nematode, *Monochamus alternatus*, is not found in the Russian region. However, six other species of *Monochamus* are present in Russia. Four species of *Monochamus* (*M. galloprovincialis, M. saltuarius, M. sutor* and *M. urussovi*) are widespread in Russia and western Europe. Individuals of *M. galloprovincialis* are known vectors of *B. xylophilus* in Portugal (Sousa *et al.*, 2001). *Monochamus impulviatus* also occurs in Russia but is rare and its distribution is limited to the Ural Mountains and along the Pacific coast. *Monochamus nitens* and *M. saltuarius* are known vectors of *B. xylophilus* in Japan and China. Although *M. nitens* is considered a dangerous pest to European forests, it occurs only on the southern end of Sakhalin Island located north of Japan. Each of the six species of *Monochamus*, with the exception of *M. nitens*, can be found in the different climatic regions of Russia.

Acknowledgment

We thank Dr Dale R. Bergdahl and Shari Halik, Department of Forestry, University of Vermont, Burlington, Vermont, USA, for their constructive comments and help with the English. This work was supported by the Foundation for Fundamental Research, the Russian Academy of Sciences.

References

BRAASCH, H. (1994). Untersuchungen zur Pathogenitat der Kiefernholznematoden (*Bursaphelenchus xylophilus*) fur verschiedene Koniferenarten unter mitteleuropaischen Freilanbedingungen. *Mitteilungen aus der Biologischen Bundesanstadt für Land- und Forstwirtschaft Berlin-Dahlem* 301.

BRAASCH, H., TOMICZEK, C., METGE K., HOYER, U., BURGERMEISTER, W., WULFERT, I. & SCHÖNFELD, U. (2001). Record of *Bursaphelenchus* spp. (Nematoda, Parasitaphelenchidae) in coniferous timber imported from the Asia part of Russia. *Forest Pathology* 31, 129-140.

EROSHENKO, A.S. & KRUGLIC, I.A. (1996). Korean pine pathogenic nematodes in the Russia Far East South. In: *The Cedar and Deciduous Forest of the Far East*, Khabarovsk, pp. 196-197.

KULINICH, O.A. (1991). [The nematodes of wood pine.] *IX Nematology Symposium, Kishinev, 1991*. Kishinev, pp. 16-17. [Abstr.]

KULINICH, O.A. & KOLOSSOVA, N.V. (1995). On the possibility of establishment of the pinewood nematode *Bursaphelenchus xylophilus* in the countries of the former USSR. *Russian Journal of Nematology* 3, 35-48.

KULINICH, O.A. & ORLINSKII, P.D. (1998a). Results of cross-hybridization of *Bursaphelenchus mucronatus* isolates from Russia and France and *B. xylophilus* isolates from Canada. In: Futai, K., Togashi, K. & Ikeda, T. (Eds). *Sustainability of pine forests in relation to pine wilt and decline. Proceedings of the symposium, Tokyo, Japan, 26-30 October 1998*. Kyoto, Japan, Shokado Shoten, pp. 24-25.

KULINICH, O.A. & ORLINSKII, P.D. (1998b). Distribution of conifer beetles (Scolytidae, Curculionidae, Cerambycidae) and wood nematodes (*Bursaphelenchus* spp.) in European and Asian Russia. *EPPO Bulletin*, 28, 39-52.

KULINICH, O.A. & ORLINSKII, P.D. (2001). Pathogenicity of the pinewood nematode *Bursaphelenchus* spp. in Central European Russia. *IVth International Symposium of the Russian Society of Nematologists, Moscow, 2001*. Moscow, 133. [Abstr.]

KULINICH, O.A., KRUGLIC, I.A., EROSHENKO, A.S. & KOLOSSOVA, N.V. (1994). Occurrence and distribution of the nematode *Bursaphelenchus mucronatus* in the Russian Far East. *Russian Journal of Nematology* 2, 113-120.

KULINICH, O.A., MAGNUSSON C. & KOLOSSOVA N.V. (1996). Notes on the comparative morphology of the isolates of *Bursaphelenchus mucronatus* from Russia. *Russian Journal of Nematology* 4, 88. [Abstr.]

KULINICH, O.A., ORLINSKII, P.D. & KOLOSSOVA, N.V. (1998). Cross-hybridization of *B. xylophilus* isolates from Canada with *B. mucronatus* isolates from Russia. *Second International Symposium of the Russian Society of Nematologists, Moscow, 23-30 August 1997. Russian Journal of Nematology* 6, 66. [Abstr.]

KULINICH, O.A., TJURIN, A.A. & ORLINSKII, P.D. (1999). Surveys for the pinewood nematode and a review of the related research in Russia. In: Futai, K., Togashi, K. & Ikeda, T. (Eds). *Sustainability of pine forests in relation to pine wilt and decline. Proceedings of the symposium, Tokyo, Japan, 26-30 October 1998*. Kyoto, Japan, Shokado Shoten, pp. 47-51.

LINIT, M.J. (1988). Nematode-vector relationships in the pine wilt disease system. *Journal of Nematology* 20, 227-235.

LINIT, M.J. (1990). Transmission of pinewood nematode through feeding wounds of *Monochamus carolinensis* (Coleoptera: Cerambycidae). *Journal of Nematology* 22, 231-236.

MAGNUSSON, C. & KULINICH, O.A. (1996). A taxonomic appraisal of the original description, morphology and status of *Bursaphelenchus kolymensis* Korentchenko, 1980 (Aphelenchida: Aphelenchoididae). *Russian Journal of Nematology* 4, 155-161.

MAMIYA Y. (1999). Review on the pathogenicity of *Bursaphelenchus mucronatus.* In: Futai, K., Togashi, K. & Ikeda, T. (Eds). *Sustainability of pine forests in relation to pine wilt and decline. Proceedings of the symposium, Tokyo, Japan, 26-30 October 1998.* Kyoto, Japan, Shokado Shoten, pp. 57-64.

PANESAR, T.S. & SUTHERLAND, J.R. (1989). Pathogenicity of Canadian isolates of the *Bursaphelenchus xylophilus* (pinewood nematode) to provenances of *Pinus sylvestris* and *Pinus contorta* as grown in Finland: a greenhouse study. *Scandinavian Journal of Forage Research* 4, 549-557.

SOUSA E., BRAVO M.A, PIRES, J., NAVES, P., PENAS, A.C., BONIFACIO, L. & MOTA, M.M. (2001). *Bursaphelenchus xylophilus* (Nematoda: Aphelenchoididae) associated with *Monochamus galloprovincialis* (Coleoptera: Cerambicidae) in Portugal. *Nematology* 1, 89-91.

Nematology Monographs & Perspectives, 2003, Vol. 1, 77-91

The pine wood nematode problem in Europe – present situation and outlook

Helen BRAASCH [1] and Siegfried ENZIAN [2]

[1] *Department for National and International Plant Health and*
[2] *Institute for Technology Assessment in Plant Protection,*
Federal Biological Research Centre for Agriculture and Forestry,
Kleinmachnow Branch, Stahnsdorfer Damm 81, 14532 Kleinmachnow,
Germany

Summary – The detection of the quarantine pest *B. xylophilus* in Portugal in 1999 has indicated the need to know more about the distribution of *Bursaphelenchus* spp. in coniferous trees in Europe in order to establish the true geographic range of the species and to act quickly in case of its unwanted introduction into other European regions. The results of the surveys on the occurrence of *Bursaphelenchus* species executed in southern and central Europe in recent years and of the monitoring for *B. xylophilus* in the EU member states in 2000 are presented. They contributed to the knowledge of the distribution of the genus *Bursaphelenchus* in Europe. The dangerous pine wood nematode was not found outside Portugal and also not outside the infested area on the Setúbal Peninsula in Portugal. Commission Decisions of the European Union in 2000 and 2001 provide temporary measures for the member states against the dissemination of *B. xylophilus* to uninfested areas in Portugal and to other countries. Portugal continues to take measures to control the spread of the pest and to eradicate it. A further EU decision of 2001 provides temporary emergency measures in respect to wooden packaging material of coniferous wood originating in Canada, China, Japan and USA in order to prevent the introduction of the pine wood nematode, because it has repeatedly been intercepted in packaging wood from these countries. The monitoring survey in EU member states will be continued, and other European countries will do likewise. In order to achieve reliable results, the monitoring has to be focused on places and regions particularly endangered by introduction and establishment of the nematode. By combining data on distribution of susceptible pine species and climatic conditions, the potential threat of a further spread of the pine wood nematode in Europe is estimated.

A pest risk analysis (PRA) for the territories of the European Union was carried out on *Bursaphelenchus xylophilus* (Steiner & Buhrer, 1934) Nickle, 1970 (pine wood nematode) and its vectors in the genus *Monochamus* by an EU expert group a few years ago (Evans *et al.*, 1996). Taking account of the fact that *B. xylophilus* did not occur in the PRA area at that time, the key conclusions of this PRA were that the entire PRA area would be suitable for establishment of the nematode, susceptible host species occur universally in the PRA area, economic important damage to these hosts could be expected in some regions and the presence of *B. xylophilus* would be prejudicial to Community trade. The conclusion was that all commodities of coniferous wood from infested areas/sources warrant phytosanitary measures. A detailed analysis was made of the risks presented by different trade pathways and of the phytosanitary measures to reduce the risk.

Bursaphelenchus xylophilus is a quarantine organism for the European Union (Directive 2000/29 EC, former Directive 77/99/EEC). Annex 4A of this directive relating to plants and measures against pests lays down *inter alia* the specific requirements for the import of coniferous wood, specified according to botanical species, origin and wood commodity. Wood of conifers except *Thuja*, other than in the form of chips, packing cases, dunnage *etc.,* but including that which has not kept its natural round surface, must be heat-treated, reaching a core temperature of at least 56°C for a period of 30 min. Wood in the form of chips *etc.* must be heat-treated or undergo an appropriate fumigation. Wood of *Thuja* originating in areas where the pine wood nematode occurs is required to be stripped of its bark and to be free from grub holes of the genus *Monochamus*; coniferous packaging wood must additionally be kiln-dried to below 20% moisture content.

The situation in Europe has changed since the pine wood nematode was detected in Portugal in 1999 (Mota *et al.*, 1999). First of all, the area affected by *B. xylophilus* in Portugal had to be clearly defined. Only trees of *Pinus pinaster* were found infested with *B. xylophilus*. The areas in Portugal where the pine wood nematode is present are several counties in the districts Évora, Santarém and Setúbal. *Monochamus galloprovincialis* was found to be the vector of *B. xylophilus* in Portugal (Sousa *et al.*, 2001). It has not yet been possible to identify the source of contamination responsible for the establishment of the nematode in Portugal, although elements indicate that packaging material is the most likely pathway. The Portuguese authorities were aware of the seriousness of the situation and a campaign (PROLUNP) was started in order to eradicate *B. xylophilus* from Portugal. This included sub-programmes on survey, eradication and research, and a close surveillance and monitoring of movements of wood material. About 50 000 pine trees suspected to be attacked by *B. xylophilus* were felled in Portugal in 2000.

Commission Decisions of the EU in 2000 and 2001 (2000/58/EC and 2001/218/EC) set out measures that should be taken in Portugal in order to eradicate the nematode and to prevent its spread, as well as emergency measures to prevent its introduction into other territories of the EU. EU missions to Portugal were carried out under the general provisions of Community legislation and, in particular, of Council Directive 2000/29/EC, in order to check the implementation of Commission Decisions and to assess the evolution of the situation regarding outbreaks of the pine wood nematode. This changed situation clearly indicates that an updated pest risk assessment for Europe is urgently required.

This paper explains the Commission Decisions with regard to the *B. xylophilus* outbreak in Portugal, the results of the year 2000 monitoring survey of *B. xylophilus* in the EU territories, and preliminary results relevant for a new PRA for Europe based on distribution of susceptible pine species and climatic data.

Material and methods

The Commission Decisions 2000/58/EC of 11 January 2000, 2001/218/EC of 12 March 2001 and 2001/219/EC of 12 March 2001 are presented in summary in regard to the measures provided in their annexes. The results of the official surveys carried out by the member states in 2000 correspond to the data provided to the member states by the Standing Committee on Plant Health. They are supplemented by the results of an EU research project concluded in 2000 (Braasch *et al.*, 2000) and by surveys in non-EU countries.

An attempt was made to achieve a more detailed PRA by combining the distribution of susceptible European pine species and climatic data using GIS system. The pine species taken into consideration were the susceptible species *P. pinaster*, *P. sylvestris* and *P. nigra*. The

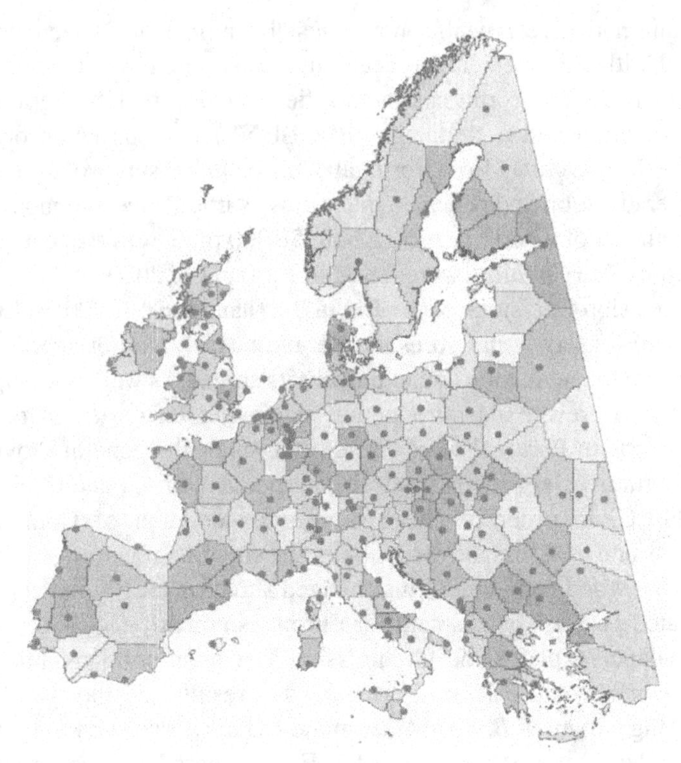

Fig. 1. *Weather stations (●) and assigned regions.*

databases of GIS application were monthly long-term means (30 years) of temperature and precipitation from 192 weather stations in Europe (DWD, 1994) and data about the distribution of pines in a pixel-width of 16 × 16 km (UN/ECE & EC, 2001). At first, every weather station was allocated to an area with the Euclid-distance-function by the GIS-software ArcView™. Every point (pixel) was allocated according to its minimum distance to a weather station. The result of this method can be seen in Fig. 1.

The weather data and the distribution of susceptible pines were combined. Since the hottest months of the year are the most important ones for development of pine wilt following attack by the pine wood nematode, the mean July/August temperature was used. The preliminary results should be considered as a first attempt to classify risk areas in Europe.

Results

REGULATIONS

Portugal informed the other member states and the Commission of the European Union on 25 June 1999 that pine tree samples originating in its territory (the Setúbal Peninsula) were infested by *B. xylophilus*. It was necessary for Portugal to take specific measures against this quarantine pest and for the other member states to adopt additional measures to protect themselves against that danger. Sweden, on the basis of this information, adopted on 29 September 1999 certain additional measures, including a special heat treatment and the use of a plant passport for all wood leaving Portugal, with a view to strengthening protection against the introduction of the pine wood nematode from Portugal. The EU Commission, by Decision 2000/58/EC of 11 January 2000, authorised member states temporarily to take additional measures against the dissemination of the pine wood nematode. Such measures should not be applied to wood from areas in Portugal in which *B. xylophilus* is known not to occur and to wood of *Thuja*. Portugal should ensure that the conditions laid down in the annex to this decision be met in relation to susceptible wood, bark and plants (*Abies, Cedrus, Larix, Picea, Pinus, Pseudotsuga, Tsuga*). These measures refer to movement of wood, isolated bark and host plants from the infested area in Portugal to other member states, to movements within Portugal and to control measures against *B. xylophilus* with the aim of eradication. Member states had to conduct official surveys for *B. xylophilus* in their countries and may subject consignments of wood and bark from Portugal to testing for the presence of *B. xylophilus*.

The measures laid down in the Annex of the Decision 2000/58/EC cover inspection and use of a plant passport for plants, heat treatment and plant passports for wood and isolated bark, fumigation and plant passports for chips *etc.*, bark stripping, freedom from grubholes and moisture content below 20% for packaging wood with regard to movements into other member states. Movements within Portugal are differentiated into movements between 1 November and 1 March and movements between 2 March and 31 October. Trees with *B. xylophilus* or showing any symptoms of poor health had to be felled and immediately destroyed (or appropriately processed) within the infested area in the summer period, whereas in the winter period, felled trees from the infested area could also be moved, after stripping bark and under official control, to a processing plant anywhere in Portugal, where the wood had to be appropriately processed or heat-treated before 2 March. Isolated bark from the infested area had to be destroyed or heat-treated. The effect of the emergency measures has been assessed continually. The results of the survey were summarised in October 2000. The Decision had to be reviewed by November 2000 at the latest.

The Commission Decision 2001/218/EC of 12 March 2001 is a revised continuation of the protective measures laid down in Decision 2000/58/EC including conduct of official surveys for the pine wood nematode in 2001. The effect of the measures will be assessed during 2001 and 2002. In contrast to the former decision, which considers three types of areas in Portugal (infested zone, buffer zone and area where *B. xylophilus* is known not to occur) the year 2001 decision defines only two types of areas in Portugal: The so-called demarcated areas comprise a part in which *B. xylophilus* is known to occur and a part designated as buffer zone of not less than 20 km radius surrounding that part, taking into account the results of the surveys. Portugal shall establish a second type of area in which *B. xylophilus* is known not to occur, and the Commission will compile a list of these areas for the information of the member states. Provisions referred to in the annex concern the movement of plants, wood and bark from demarcated areas into areas in Portugal other than demarcated areas, or into other member states, and cases of movement within demarcated areas of Portugal. Due to the results of vector research (hatching and flight period) in Portugal, the winter period was extended to the period 1 November to 1 April. The measures against the spread of the pine wood nematode from the demarcated area into new regions include inspection (plants), use of plant passports, heat treatment (wood and isolated bark), fumigation (chips *etc.*) as well as bark stripping, freedom from grubholes and moisture content below 20% (dunnage *etc.*). Heat treatment, pressure impregnation or fumigation as well as marking or plant passports are used for packaging wood. Member states other than Portugal may subject consignments of susceptible wood, bark and plants coming from demarcated areas in Portugal to testing for the presence of *B. xylophilus*.

Wooden packaging is the perfect pathway for the introduction of pests. Packaging material is often manufactured of freshly cut raw wood, often includes bark, and is often made of wood whose quality is classed as low because of pest damage. Infestations of imported non-manufactured coniferous wood packaging material originating in Canada, the United States of America, Japan and China were reported by Finland, Sweden and France in 2000. In addition, *Monochamus* spp., vectors of *Bursaphelenchus*, have repeatedly been found in imported coniferous packaging wood. This shows that previous measures concerning coniferous packaging wood were not sufficient to protect the Community adequately against the introduction of *B. xylophilus*.

The Commission Decision 2001/219/EC authorised member states temporarily to take emergency measures in respect of wood packaging of non-manufactured coniferous wood originating in Canada, China, Japan and the United States of America. The measures had to be adopted by the member states by 30 September 2001 at the latest. The emergency measures should be applied in two steps. In the first step, the member states should immediately carry out an official monitoring of this wood in order to further reduce the risk. The provisions of this Decision referred to in the Annex are the second step. Susceptible wood originating in Canada, the United States or Japan shall be heat treated, pressure impregnated or fumigated and display an officially approved marking. Susceptible wood originating in China shall also be subject to one of these measures and must be accompanied by a certificate (for all regions of provenance), but the measures are not applicable to wood originating in areas in China in which *B. xylophilus* is known not to occur. If the provisions have not been complied with, the member states shall ensure that the wood be appropriately treated, destroyed or processed.

Monitoring survey

The Commission Decision 2000/58/EC obliged the member states to conduct official surveys for the pine wood nematode on susceptible trees, wood and bark originating in their countries to determine whether there is any evidence of infestation by this pest. An EU expert group prepared guidelines for the procedure (EC, Pine wood nematode survey, Protocol 2000). The results were notified to the Commission and to the other member states.

Altogether, 5200 samples (mixed wood samples of one to five conifer trees) were investigated in the EU countries in summer 2000. Especially extensive sampling was done in Spain (1757 samples), Finland (1000 samples), Portugal (611 samples), Germany (355 samples) and France

Table 1. *Possible criteria for risk classes for potential development of pine wilt based on long-term mean July/August temperature and long-term annual precipitation and the presence of susceptible pine species.*

Risk class	Temperature (°C)	Precipitation (mm)
Extreme	>25	<500 (600)
High	20-25	<500 (600)
	>25	>500 (600)
Moderate	18-20	<500 (600)
	20-25	>500 (600)
Low	<18	<> 500 (600)

(307 samples). Additionally, 1250 mixed samples (each of ten trees) had been checked in Germany, Greece, Italy and Austria in the frame of an EU research project (RISKBURS) in 1996-1999 (Caroppo *et al.*, 1998; Braasch *et al.*, 1999; Skarmoutsos & Skarmoutsos, 1999; Braasch *et al.*, 2000; Tomiczek, 2000). Among the 5200 samples taken mainly from damaged trees of *Pinus pinaster, P. sylvestris* and *P. nigra*, 1540 samples originated from risk areas such as places around harbours, airports, timber yards, saw mills, pulping plants, *etc.* In addition, branches left in clear cutting areas were sampled. Neither in the RISKBURS sampling nor in the 2000 survey was *B. xylophilus* found outside Portugal. In Portugal, there was only a minor extension of the infested area into the buffer zone in a southerly direction. In spite of intensive tree fellings and other measures under the eradication programme, the phytosanitary situation in Portugal has improved only slightly. Among the ten other *Bursaphelenchus* spp. of which the Commission was notified in 2000, *B. mucronatus* and *B. sexdentati* were the most frequently recorded species.

Monitoring surveys for the pine wood nematode have also been conducted in other European countries (non-EU). Extensive sampling was, for instance, done in Norway (McNamara & Stoen, 1988; Magnusson *et al.*, 2001), Poland (Brzeski & Brzeski, 1997), Cyprus (Braasch & Philis, 2002) and Russia, both Asian and European parts (Kulinich & Orlinskii, 1998). No indications exist that *B. xylophilus* is present in any other European region except Portugal.

Fig. 2. *Map of the distribution of susceptible* Pinus nigra, P. pinaster *and* P. sylvestris *(■, 16 km grid) spp. in Europe (except eastern part). (Source UN/ECE & EC, 2001.)*

PEST RISK ANALYSIS

Pine forests seem to be vulnerable if the temperature is high enough (more than 20°C) for long enough (at least 8 weeks) and providing that susceptible tree species are present. Several abiotic and biotic factors are reported to predispose trees to infection, among them water stress. The factors of summer temperature and annual precipitation were connected with the distribution of susceptible pine species in order to show the regions most at risk for development of pine wilt disease. Long-term data for July/August temperature and annual precipitation and the distribution of susceptible pines are indicated in Figs 2-4. These data are combined in Figs 5 and 6, and preliminary risk classes of 'extreme', 'high', 'moderate' and 'low' were established in order to demonstrate differences between regions. These terms depend, of course, on the data limits selected for the risk classes. For instance, the more highly endangered regions would extend farther north if 600 mm precipitation

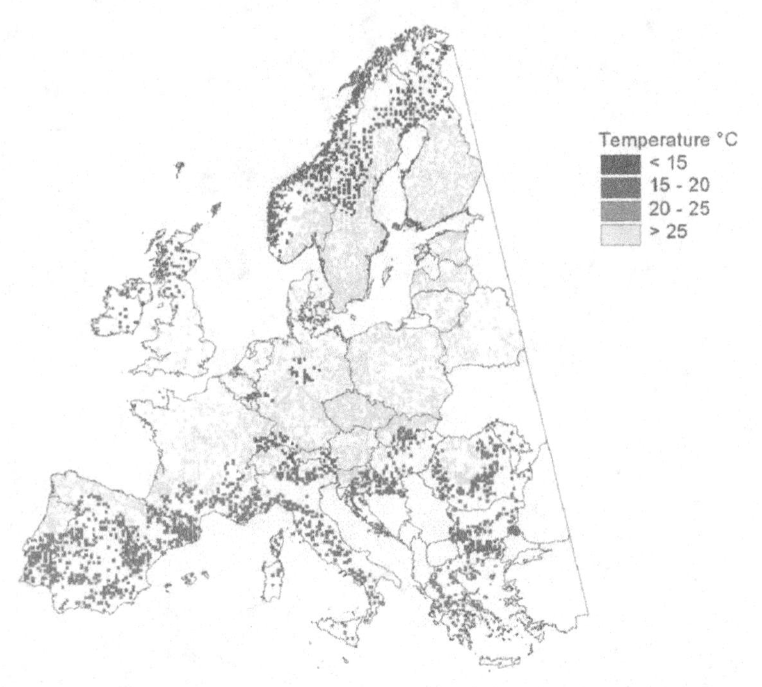

Fig. 3. *Map of long-term means of July/August temperature in Europe (except eastern part).*

had been selected (Fig. 6) instead of 500 mm (Fig. 5) as a critical amount. The classification shown in Table 1 is an attempt at establishing a possible procedure.

The results of the procedure show that, indeed, southern Europe is the most endangered region. If 600 mm precipitation is taken as the critical amount of precipitation, a moderate danger is estimated for eastern central Europe including the very east of Germany as well. As mentioned above, the procedure relies on long-term data. In successive years with relatively high summer temperature the risk may increase in the affected regions. An accelerated development of vector populations under these conditions may contribute to this situation.

Discussion

Although *B. xylophilus* is not considered an important pest of native pines in North America, it is responsible for causing significant pine

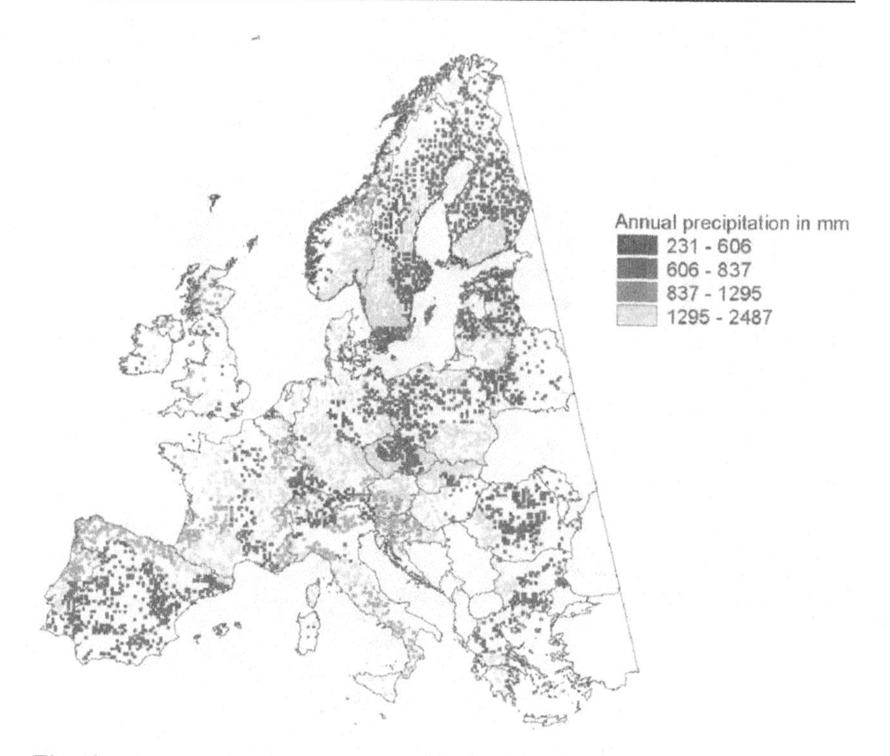

Fig. 4. *Map of long-term means of annual precipitation in Europe (except eastern part).*

wilt disease in Japan, China, South Korea and Taiwan. Its establishment in Europe and the damage caused in Portugal confirm the conclusions of the PRA of the EU expert group that the pine wood nematode posed a potential threat to European forests and relevant industries (Evans *et al.*, 1996). The danger linked to the international transport of coniferous packaging wood was assessed as 'relatively high' and 'similar to that for sawn wood'. The quarantine requirements for packaging wood laid down in Annex IV of the previous version of EU Directive 2000/29/EC (*i.e.*, Directive 77/93/EEC) did not fully reflect this statement and were less than for sawn wood. Most probably, the pine wood nematode was introduced into Europe on this pathway. Therefore, the EU emergency measures in respect of wood packaging are highly justified.

Although about 50 000 trees were cut in the infested area in Portugal in 2000, *B. xylophilus* was detected in the southern part of the buffer zone that surrounds the infected zone in April 2001. In 2001, cutting of trees in the infected area reached about the same extent as in the

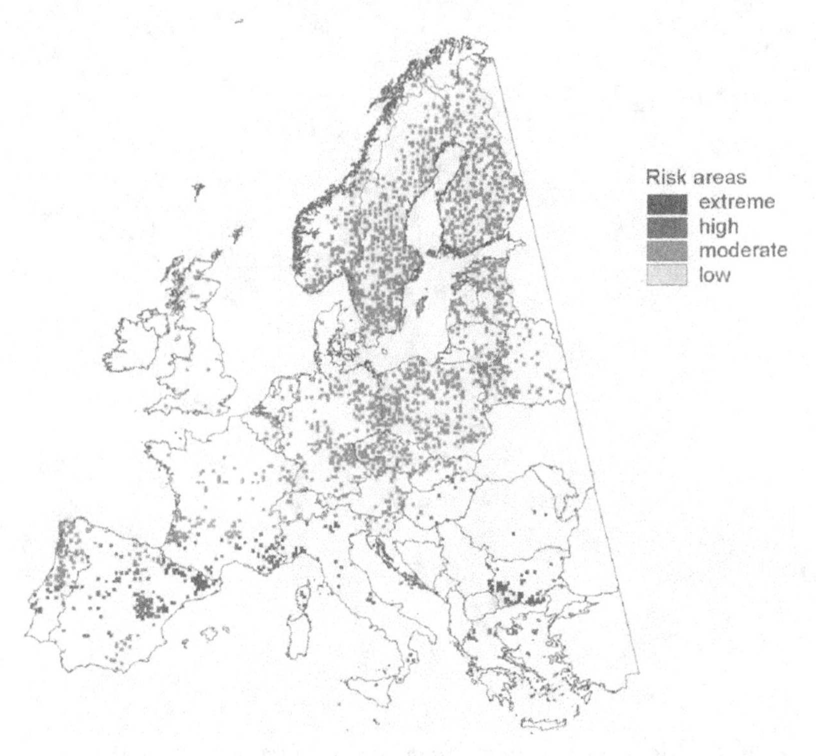

Fig. 5. *Map of regions most endangered by the pine wood nematode in Europe (except eastern part, classification in Table 1, precipitation/year <>500 mm).*

previous year. The success of the measures foreseen in EU Decisions 2001/218/EC and 2000/58/EC may be seen in the exclusion of the nematode from other countries. No pine wood nematode has been found in the member states outside Portugal and outside the demarcated area in Portugal in the monitoring surveys conducted so far. The main forest region of Portugal in the northern part of the country is free from the pest. The efforts to eradicate the nematode in Portugal and to prevent its spread to other regions need, however, to be strengthened. If no positive results are achieved, the use of insecticides for vector control as practised in Japan should be considered. Although no pine wood nematode has been found in member states outside Portugal, the official surveys in these member states also need to be intensified in order to detect as early as possible any further occurrence of *B. xylophilus*. Sampling should be biologically adequate, including vector sampling, and should concentrate on geographic and climatic risk areas.

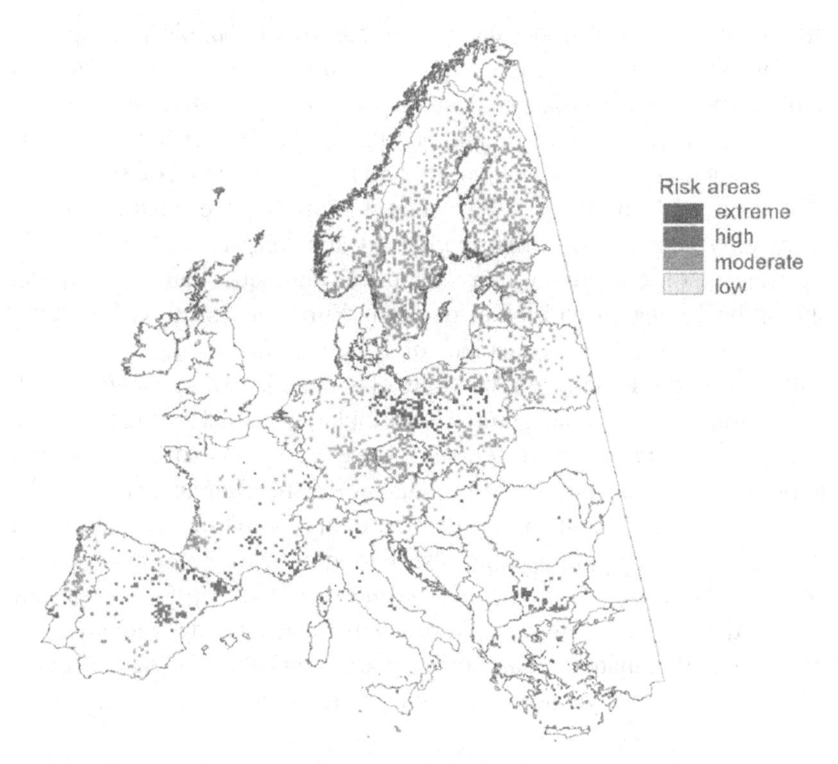

Fig. 6. *Map of regions most endangered by the pine wood nematode in Europe (except eastern part, classification in Table 1, precipitation/year <>600 mm).*

A pest risk assessment has to consider many aspects. Beside distribution, detection and taxonomic affiliation of the pest, presence of suitable vectors, biological characteristics of the pest and its vectors, adaptability and potential of the pest for establishment, the danger of the pest and its spread, dissemination and pathway analysis, the presence of suitable hosts and climatic conditions are key factors. One of the conclusions of the PRA for the European Union from 1996 (Evans *et al.*, 1996) was that the entire PRA area is suitable for establishment of *B. xylophilus* and pine wilt can be expected in the Mediterranean and continental dry regions. It is obvious that regions endangered by pine wilt disease deserve an especially careful monitoring survey because they have optimal conditions for establishment of the pest and for symptom expression, and they may serve as centres of further distribution of the pine wood nematode, due to the breeding availability of the vectors and the high multiplication rate of the nematode. In spite of favourable climatic conditions in southern Europe, the distribution of susceptible pine species and of the vector beetles is unequal. The distribution of *Monochamus* spp. seems to be approximately linked to the distribution of conifers, whereas their real occurrence and population density depend on various factors, such as breeding availability and annual climatic conditions.

A new and improved PRA should consider the distribution of susceptible host plants, distribution of the vectors and geographic differences of climate in more detail. Ecoclimatic data of Portugal should be compared with other places in Europe by using GIS system. An updated pest risk assessment of *B. xylophilus* for Europe, which will permit predictions on the rate of spread of *B. xylophilus* with and without active management and includes model calculations on the effect of various management strategies, will be of fundamental importance for developing new quarantine regulations by the EU. The preliminary results presented here of combining climatic data (temperature, precipitation) and distribution of susceptible pines (*P. pinaster, P. sylvestris, P. nigra*) are considered a first step towards a new PRA and are aimed at giving an idea where the most endangered regions to pine wood nematode attack in Europe (except the very east) may be found. They have to be improved by using regionally refined data.

References

BRAASCH, H. & PHILIS, J. (2002). New records of *Bursaphelenchus* spp. in Cyprus. *Nematologia Mediterranea* 30, 55-57.

BRAASCH, H., METGE, K. & BURGERMEISTER, W. (1999). [*Bursaphelenchus* species (Nematoda, Parasitaphelenchidae) found in coniferous trees in Germany and their ITS-RFLP patterns.] *Nachrichtenblatt des Deutschen Pflanzenschutzdienstes* 51, 312-320.

BRAASCH, H., BURGERMEISTER, W., HARMEY, M.A., MICHALOPOULOS-SKARMOUTSOS, H., TOMICZEK, C. & CAROPPO, S. (2000). *Pest risk analysis of pinewood nematode related* Bursaphelenchus *species in view of South European pine wilting and wood imports from Asia.* Final Report of EU research project FAIR CT 95-0083, 251 pp.

BRZESKI, M.W. & BRZESKI, J. (1997). Survey of *Bursaphelenchus* (Nematoda: Appelenchoididae) species in pine wood in Poland. *Fragmenta Faunistica* 40, 103-109.

CAROPPO, S., AMBROGIONI, L., CAVALLI, M. & CONIGLIO, D. (1998). Investigations on the occurrence of *Bursaphelenchus* nematodes in coniferous trees and their possible vectors in Italy. *Nematologia Mediterranea* 26, 97-116.

DWD DEUTSCHER WETTERDIENST (1994). *Klimadaten ausgewählter europäischer Stationen.*

EVANS, H.F., MCNAMARA, D.G., BRAASCH, H., CHADOEUF, J. & MAGNUSSON, C. (1996). Pest Risk Analysis (PRA) for the territories of the European Union (as PRA area) on *Bursaphelenchus xylophilus* and its vectors in the genus *Monochamus*. *EPPO Bulletin* 26, 199-249.

KULINICH, O.A. & ORLINSKII, P.D. (1998). Distribution of conifer beetles (Scolytidae, Curculionidae, Cerambycidae) and wood nematodes (*Bursaphelenchus* spp.) in European and Asian Russia. *EPPO Bulletin* 28, 39-52.

MAGNUSSON, C., THUNES, K., HAUKELAND SALINA, S. & OEKLAND, B. (2001). Survey of the pine wood nematode, *Bursaphelenchus xylophilus*, in Norway 2000. *Planteforsk rapport* 07/2001, pp. 1-20.

MCNAMARA, D.G. & STOEN, M. (1988). A survey for *Bursaphelenchus* spp. in pine forests in Norway. *EPPO Bulletin* 18, 353-363.

MOTA, M.M., BRAASCH, H., BRAVO, M.A., PENAS, A.C., BURGERMEISTER, W., METGE, K. & SOUSA, E. (1999). First record of *Bursaphelenchus xylophilus* in Portugal and in Europe. *Nematology* 1, 727-734.

SKARMOUTSOS, G. & SKARMOUTSOS, H. (1999). First record of *Bursaphelenchus* nematodes from pine forests in Greece. *Plant Disease* 83, 879.

SOUSA, E., BRAVO, M.A., PIRES, J., NAVES, P., PENAS, A.C., BONIFACIO, L. & MOTA, M.M. (2001). *Bursaphelenchus xylophilus* (Nematoda; Aphelenchoididae) associated with *Monochamus galloprovincialis* (Coleoptera; Cerambycidae) in Portugal. *Nematology* 3, 89-91.

TOMICZEK, C. (2000). A survey for *Bursaphelenchus* spp. in conifers in Austria and implications to Austria forests. *XXI IUFRO World Congress 2000, 7-12 August, Kuala Lumpur, Malaysia*, p. 399. [Abstr.]

UN/ECE & EC (2001). *Forest Condition in Europe.* Report on the 2000 Survey. Geneva, Brussels, United Nations, European Commission, 103 pp.

Nematology Monographs & Perspectives, 2003, Vol. 1, 93-99

The genus *Bursaphelenchus* (Nematoda) in Spain

Miguel ESCUER, Maria ARIAS and Antonio BELLO

Departamento Agroecología, CCMA, CSIC, Serrano 115 dpdo, 28006 Madrid, Spain

Summary – A survey to study nematodes from the family Aphelenchida is being carried out in Spain, because of the report of *Bursaphelenchus xylophilus*, pine wood nematode (PWN), in Portugal, and following a directive from the EU. Samples were taken, coordinated by the Spanish Ministry of Agriculture, Fisheries and Food, with the cooperation of the official forestry and plant health services and diagnostic laboratories of the different autonomous communities, mainly in pine forest close to the Portuguese border, in sawmills and also in wood imported *via* the responsible Customs and Excise authority. Some 2794 samples were taken and 291 of these were sent to our laboratory to be studied: 160 were from dead, wilting or symptomless pine trees, or from sawmills, and 131 from imported wood. Four species of *Bursaphelenchus* were found: *B. fungivorus* in a pine wood sawmill in the Cazorla mountains in Andalucia; *B. mucronatus* associated with *Pinus halepensis, P. nigra* and *P. sylvestris* in Navarra province, as well as in importing wood mills from Asturias and Galicia; *B. pinasteri* on *P. pinaster* in Extremadura, and *B. sexdentati* on *Abies alba*, *P. pinaster* and *P. pinea* in Galicia, Madrid, Navarra, Basque Country and Valencia, and on sawmills from Asturias.

Bursaphelenchus xylophilus and its vector insect were listed as A1 quarantine pests by EPPO (1986) and again in 1992 in the Annex IV of Directive 17/93/CEE to prevent the nematode's introduction from non-European countries by establishing Customs regulations for conifer wood importation. Evans *et al.* (1996) made a detailed analysis of the risk of *B. xylophilus* to the European Union. However, in May 1999, this species was reported in Setúbal Peninsula (Portugal) where it seems to be well established (Mota *et al.*, 1999; Sousa *et al.*, 2001). A survey was made in Spain following EPPO recommendations for preventing the introduction of this nematode into countries where it does not occur and according to article 3 in the 2000/58/EC Commission Decision.

The Spanish Ministry of Agriculture, Fisheries and Food established a cooperation between the official forestry services and the diagnostic laboratories of the different autonomous communities to carry out the survey. The Nematological Laboratory at the Department of Agroecology, CCMA, CSIC, was designated a reference centre for the purpose of training experts from various communities, to standardise sampling and extracting methods, to identify nematodes when experts from the autonomic laboratories failed to do so, to observe the *Bursaphelenchus* spp. that were found, and to analyse imported wood samples from Customs. *Bursaphelenchus mucronatus, B. pinasteri* and *B. sexdentati* were found in this programme (Abelleira *et al.*, 1999; Escuer & Bello, 2000; Escuer *et al.*, 2000, 2001). This paper deals with the distribution of *Bursaphelenchus* spp. in Spain.

Materials and methods

The forestry services made the surveys and send the samples to the diagnostic laboratories in each autonomous community, where they were processed to collect nematodes. Samples where aphelenchoid nematodes appeared or where the presence of *Bursaphelenchus* was suspected were sent to the reference laboratory to be analysed, identified or confirmed. The number of samples from each autonomous community was determined according to their woodlands, emphasising those bordering Portugal. Samples of imported wood were also submitted by Customs officials. The number of samples that were recovered from each autonomous community and from Customs and sent to the reference laboratory during years 2000 and 2001 are shown in Tables 1 and 2. Of the 2794 *Pinus* trees or sawmill samples collected in the various autonomies, 160 suspected of containing aphelenchoid nematodes were studied in our laboratory, as well as 131 from imported wood sent by Customs.

Results

The best-studied autonomous communities were those with extensive woodlands, bordering Portugal or those closest to ports. The most samples were taken in Galicia, followed by Castile & León, Valencia, Extremadura, Andalusia and Aragón. The least studied provinces were Murcia, Asturias and Cantabria. Of the samples sent to the reference

Table 1. *Surveys carried out in every autonomous community of Spain (Fig. 1).*

Community	Samples/year				Total		%
	2000		2001				
Andalusia	118	(0)*	47**	(11)*	165	(11)*	5.9
Aragón	115	(5)	34	(0)	149	(5)	5.3
Asturias	12	(3)	12	(3)	24	(6)	0.9
Balearic Islands	23	(23)	12	(0)	35	(23)	1.3
Cantabria	15	(4)	12	(0)	27	(4)	1.0
Castile-La Mancha	41	(1)	19	(0)	60	(1)	2.1
Castile & León	171	(31)	80	(0)	251	(31)	9.0
Catalonia	53	(0)	26	(0)	79	(0)	2.8
Extremadura	115	(15)	60	(4)	175	(19)	6.3
Galicia	792	(1)	514	(4)	1306	(5)	46.7
La Rioja	19	(1)	15	(0)	34	(1)	1.2
Madrid	65	(19)	41	(14)	106	(33)	3.8
Murcia	11	(1)	6	(0)	17	(1)	0.6
Navarre	17	(17)	14	(0)	31	(17)	1.1
Basque Country	60	(2)	47	(0)	107	(2)	3.8
Valencia	131	(0)	97	(1)	228	(1)	8.2
Total	1758	(123)	1036	(37)	2794	(160)	100

* Samples sent to the Department of Agroecología, CCMA, CSIC.
** Estimated samples for 2001.

Table 2. *Numbers and countries of origin of samples from the six Customs authorities in Spain studied by the reference laboratory in 2000 and 2001.*

Customs	Year/number of samples and country of origin				Total
	2000		2001		
Cadiz	3 (Brazil), 24 (USA)	24	2 (Brazil), 16 (USA)	18	45
Barcelona	4 (USA)	4	–	0	4
Granada	1 (USA), 1 (Russia)	2	–	0	2
Pontevedra	1 (South Africa), 4 (USA), 1 (Esthonia)	6	6 (USA)	6	12
Valencia	2 (Canada), 28 (USA)	30	2 (Canada), 28 (USA)	30	60
Biscay	1 (Canada), 2 (China), 1 (Japan), 1 (Taiwan), 1 (USA), 1 (Russia)	7	–	0	8
Total		73		54	131

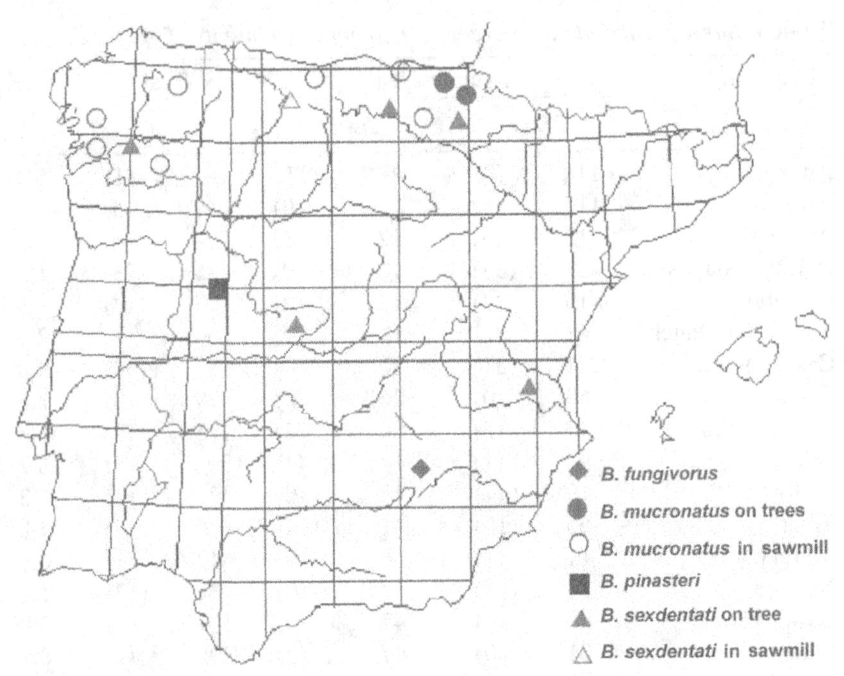

Fig. 1. Bursaphelenchus *spp. distribution in Spain.*

laboratory, only 6.8% from the year 2000 (123 samples) and only 3.5% from 2001 (37 samples) were suspected to be *Bursaphelenchus* spp.

Four species of *Bursaphelenchus* Fuchs, 1937 were found: *B. fungivorus* Franklin & Hooper, 1962; *B. mucronatus* Mamiya & Enda, 1979; *B. pinasteri* Baujard, 1980 and *B. sexdentati* Rühm, 1960 (Fig. 1).

A *B. fungivorus*-like population was found at a sawmill on *Pinus* sp. from Cazorla Mountains in Jaén (Andalusia). Its morphometric characteristics are similar to *B. hunti*, except for the presence of four lateral lines instead of three, and to *B. fungivorus*, except for body length, which was much shorter in the Andalusian population. However, according to H. Braasch (pers. comm.) it seems to belong to the latter species, because of its very wide range in body length.

B. mucronatus was found associated with *Pinus halepensis, P. nigra* and *P. sylvestris* only in the province of Navarre, northern Spain, as well as in imported wood from sawmills of Basque Country, Galicia and Cantabria.

B. pinasteri has been found on *P. pinaster* from Extremadura. Lastly, *B. sexdentati* is the most widespread species in the Spanish Peninsula.

It was found associated with *Abies alba* from Navarre, *P. pinaster* from Valencia, *P. pinea* from Madrid, *P. sylvestris* from the Basque Provinces, and in Cantabria and Galicia sawmills.

Part of these results were recovered by Braasch (2001) from our report on the 2000 survey to the Spanish Ministry of Agriculture, Fisheries and Food.

Discussion

Bursaphelenchus fungivorus has been reported from glasshouses on growing medium in Germany and on *Gardenia* sp. infected by *Botrytis cinerea* in the United Kingdom, and in coniferous bark imported from Tschenia to Germany (Braasch, 2001). It has shown a pathogenic potential to *P. sylvestris* in inoculation experiments; it caused 70% mortality and 20% partial wilting of inoculated plants (Braasch *et al.*, 1998). In Spain it has been found once on *Pinus* sp. wood from Cazorla mountains (Andalusia).

Bursaphelenchus mucronatus is widely distributed in European pine forests in Northern and Central Europe. It has been reported on *Larix decidua* from Germany, *L. sibirica* from Russia, *Picea abies* from Germany, *Pinus sylvestris* from Germany, Poland, Czech Republic, Norway, Sweden, Finland and Austria (Braasch, 2001). It is also the most common species in Italy, with distribution limited to the northern and central western areas (Ambrogioni & Caroppo, 1998) on *P. nigra austriaca, P. pinaster, P. strobus and P. sylvestris*, generally associated with the cerambycid *Monochamus galloprovincialis galloprovincialis* (Ol.), Caroppo *et al.* (1998), and has been reported from Greece on *P. brutia* (Skartmousos & Skartmousos, 1999) and on *P. pinaster* from France (Baujard *et al.*, 1979). It has been proved through experiments to be pathogenic to varying degrees, depending on the origin and virulence of the strain, to several species of *Pinus* and *Larix*, including *P. pinaster* and *P. sylvestris*. This contributes to the decline of pines in regions with low precipitation and mean temperatures in July/August between 20-25°C (Braasch, 1996; Braasch *et al.*, 1998). In Spain it has only been found associated with pine woodlands in Navarre. Other findings are from imported wood in sawmills.

Bursaphelenchus pinasteri is described on *P. pinaster* xylem from the French Landes (Baujard, 1980) and reported from *P. sylvestris* in

Germany (Braasch, 2001; Schönfeld *et al.*, 2001). It has been found once on *P. pinaster* from Extremadura and could be broadly distributed in Spain.

Finally, *B. sexdentati* has been reported as widespread in Mediterranean environments, Greece and Italy (Ambrogioni & Caroppo, 1998; Skartmousos & Skartmousos, 1999). It has also proved pathogenic in experiments on several *Pinus* spp. (Skartmousos & Michalopoulos-Skartmousos, 2000). It is widespread in the north and east of Spain on *Abies* and *Pinus* spp. and could be more widely distributed.

Acknowledgements

The authors are indebted to the MAPA Pest Service, Forestry Services, Customs officials, and technicians of the diagnostic laboratories of the autonomous communities for sampling, processing and sending samples to us. We are also grateful to Mr Casimiro Martínez for technical assistance.

References

ABELLEIRA, A., MANSILLA, J.P., AGUÍN, O., PINTOS, C. & SALINERO, M.C. (2000). Detección de especies de *Bursaphelenchus* en Galicia. *Resúmenes X Congreso de la SEF*, Valencia, Spain, 123 pp.

AMBROGIONI, L. & CAROPPO, S. (1998). Morphology and morphometrics of Italian populations of *Bursaphelenchus* species. *Nematologia Mediterranea* 26, 97-116.

BAUJARD, P. (1980). Trois nouvelles espèces de *Bursaphelenchus* (Nematoda: Tylenchida) et remarques sur le genre. *Revue de Nématologie* 5, 167-177.

BAUJARD, P., BOULBRIA, A., HAM, R., LAUMOND, C. & SCOTTO LA MASSESE, C. (1979). Premières donneés sur le nématofauna associée aux déperissement du pin maritime dans l'Ouest de la France. *Annales des Sciences Forestières* 36, 331-339.

BRAASCH, H. (1996). Pathogenitästettest mit *Bursaphelenchus mucronatus* an Kiefern und Fichtenssunlingen in Deutschland. *European Journal of Forest Pathology* 26, 205-216.

BRAASCH, H. (2001). *Bursaphelenchus* species in conifers in Europe: distribution and morphological relationships. *EPPO Bulletin* 31, 127-142.

BRAASCH, H., CAROPPO, S., AMBROGIONI, L., MICHALOPOULOS, H., SKARTMOUSOS, G. & TOMICZEK, C.H. (1998). Pathogenicity of various *Bursaphelenchus* species and implications to European forest. In: Futai, K.,

Togasmi, K. & Shokado, I.T. (Eds). *Symposium on sustainability of pine forests in relation to pine wilt and decline. Proceedings of the symposium, Tokyo, Japan, 26-30 October 1998*. Kyoto, Japan, Shokado Shoten, pp. 7-22.

CAROPPO, S., AMBROGIONI, L., CAVALLI, M. & CONIGLIO, D. (1998). Occurrence of the pine wood nematodes, *Bursaphelenchus* spp., and their possible vectors in Italy. *Nematologia Mediterranea* 26, 87-92.

ESCUER, M. & BELLO, A. (2000). Nematodos del género *Aphelenchoides* de interés fitopatológico y su distribución en España. *Boletin de Sanidad Vegetal de Plagas* 26, 47-63.

ESCUER, M., ARIAS, M. & BELLO, A. (2000). Aphelenchoidoidea nematodes of phytopathological interest and their distribution in Spain. *Nematology* 2, 754. [Abstr.]

ESCUER, M., ARIAS, A. & BELLO, A. (2001). Nematodos del orden Aphelenchida asociados a coníferas en España. *Resúmenes 33ra reunión de ONTA, Cuba, Junio 2001*, p. 29.

EVANS, H.F., MCNAMARA, D.G., BRAASCH, H., CHADOEUF, J. & MAGNUSSON, C. (1996). Pest risk analysis (PRA) for the territories of the European Union (as PRA area) on *Bursaphelenchus xylophilus* and its vectors in the genus *Monochamus*. *EPPO Bulletin* 26, 199-249.

MOTA, M.M., BRAASCH, H., BRAVO, M.A., PENAS, A.C., BURGERMEISTER, W., METGE, K. & SOUSA, E. (1999). First report of *Bursaphelenchus xylophilus* in Portugal and in Europe. *Nematology* 1, 727-734.

OEPP/EPPO (1986). Data sheets on quarantine organisms No. 158, *Bursaphelenchus xylophilus. EPPO Bulletin* 16, 55-60.

SKARTMOUSOS, G. & SKARTMOUSOS, H. (1999). First record of *Bursaphelenchus* nematodes from pine forest in Greece. *Plant Disease* 83, 879.

SKARTMOUSOS, G. & MICHALOPOULOS-SKARTMOUSOS, H. (2000). Pathogenicity of *Bursaphelenchus sexdentati, Bursaphelenchus leoni* and *Bursaphelenchus hellenicus* on European pine seedlings. *Forest Pathology* 30, 149-156.

SOUSA, E., BRAVO, M.A., PIRES, J., NAVES, P., PENAS, A.C., BONIFACIO, L. & MOTA, M. (2001). *Bursaphelenchus xylophilus* (Nematoda: Aphelenchoididae) associated with *Monochamus galloprovincialis* (Colleoptera: Cerambycidae) in Portugal. *Nematology* 3, 89-91.

Nematology Monographs & Perspectives, 2003, Vol. 1, 101-112

Survey of the pine wood nematode, *Bursaphelenchus xylophilus*, in Norway in 2000

Christer MAGNUSSON [1], Karl H. THUNES [2],
Solveig HAUKELAND SALINAS [1] and Bjørn ØKLAND [3]

[1] *The Norwegian Crop Research Institute, Plant Protection Centre, Høgskoleveien 7, N-1432 Aas, Norway*
[2] *Norwegian Forest Research Institute, Fanaflaten 4, N-5244 Fana, Norway*
[3] *Norwegian Forest Research Institute, Division of Ecology, Høgskoleveien 12, N-1432 Aas, Norway*

Summary – Two zone sites, *i.e.*, two circular areas of 50 km radius, were established in southern Norway. The zone sites were centred in Tofte (the location of a major pulp mill) and in Drammen (the site of a major timber yard). From June to October 2000, 66 forest blocks were visited, 65 of which were situated within the zone site areas. Samples were collected from 40 forest blocks, especially from wood attacked by wood boring insects. Some samples were also taken from a wood chip pile and from imported wood material. The total number of wood samples analysed for nematodes was 275. Out of these, 214 samples were collected from forest trees, stumps, timber and logging waste of *Pinus sylvestris* and *Picea abies.* Three samples contained nematodes belonging to the genus *Bursaphelenchus*, but the pine wood nematode (PWN), *Bursaphelenchus xylophilus*, was not detected. Similarly, this nematode was not detected in the ten samples of wood chips, nor in the 25 samples of imported lumber nor in the 26 samples of imported solid wood packing material.

The establishment of PWN in Norway would have an immediate severe effect on wood trade, and could possibly have negative effects also on forest health. Because of this there is an urgent need to detect a possible occurrence of this nematode and to eradicate such an infestation as soon as possible. This is of special importance in Norway due to the obvious difficulties in controlling larger infestations in an often rough and mountainous terrain.

The recent detection of the pine wood nematode (PWN), *Bursaphelenchus xylophilus*, in Portugal has changed the earlier view of Europe as an area free from this pest (Evans *et al.*, 1996). The Standing Committee on Plant Health of the European Union (EU) has reached a decision obliging each member state to conduct a survey of their territories for PWN (Anon., 2000). This paper reports on the Norwegian survey results for the year 2000.

Materials and methods

Two circular areas of 50 km radius, centred on points of exposure to wood import materials (*i.e.*, zone sites) were established in southern Norway. One zone site (A) was centred on Tofte, the location of a major pulp mill, and the other zone site (B) was centred on a major timber yard facility in Drammen (Fig. 1).

Within these zone sites, 65 forest blocks were inspected in the period 9 June to 5 October 2000. In addition to this, one block (Flisberget) was located in the province of Hedmark in eastern Norway. With regard to zone site forest blocks, 53 blocks were inspected in the province of Buskerud, while the number of visited logging sites in the provinces of Akershus and Vestfold were nine and three. In total, 12 blocks were rejected, leaving 53 for sampling and/or provision with trap-logs for *Monochamus* spp. All the forest blocks selected had been logged 1-4 years earlier.

The highest sampling intensity was located to the north and west of the Oslo fjord and to the north of lake Tyrifjord (Fig. 1). Samples from *Pinus sylvestris* and *Picea abies* were collected from 40 forest blocks. The main part of the samples (91%) was taken from trees and wood showing activity of wood boring insects. Each sample consisted of at least 500-1000 ml wood shavings, obtained by a portable BOSCH GSR 12 VE-2 reversible electric drilling machine fitted with a 20 mm diam. spiral drill. Each sample was put into a plastic bag and transported to the laboratory.

In addition to sampling forest trees and wood, the sampling also included softwood chips at Tofte pulp mill, as well as lumber and solid wood packing material imported to Norway.

In the laboratory, all wood samples were incubated at 25°C in their plastic bags for 3 weeks prior to extraction in Baermann funnels. After

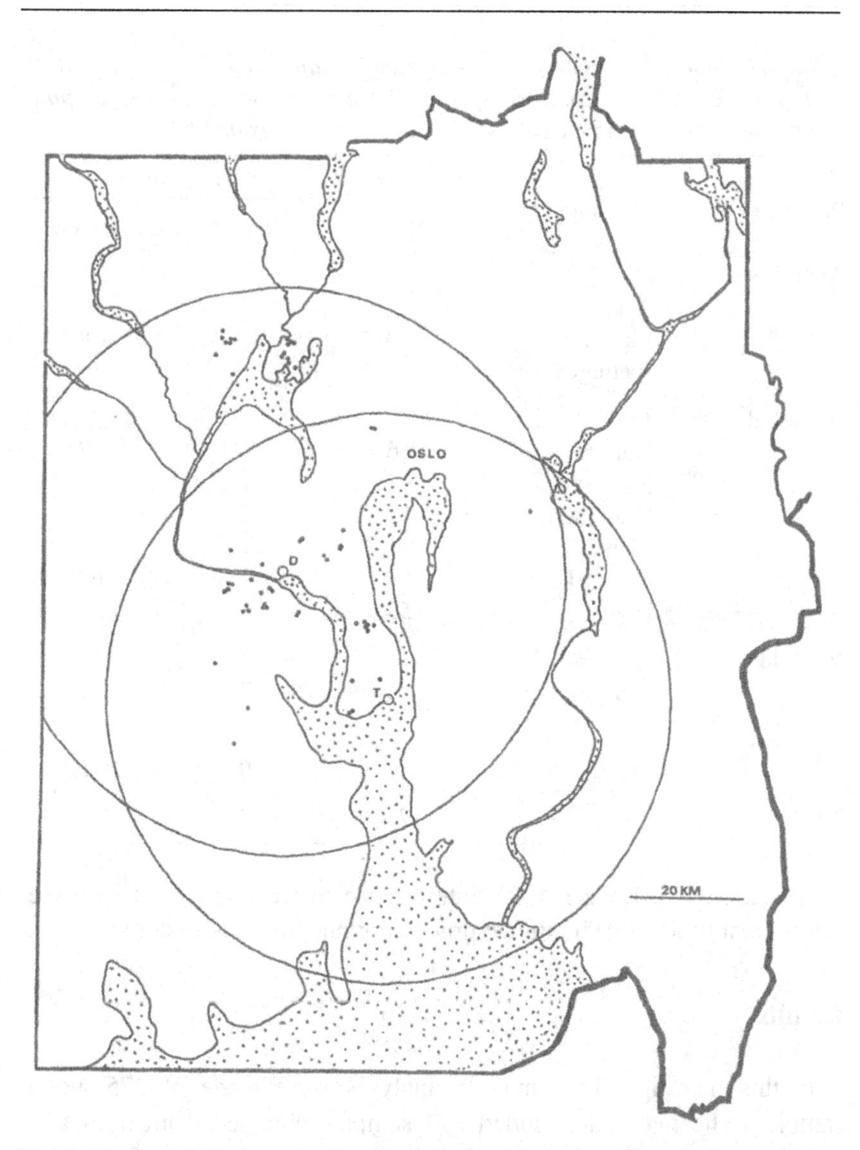

Fig. 1. *Survey of the pine wood nematode (PWN),* Bursaphelenchus xylophilus, *in Norway 2000. Location of forest blocks sampled within two zone sites of 50 km radius centred in Tofte (T) and Drammen (D).*

24 h the water was removed and the nematode suspension was allowed to settle. The nematodes were killed at 65°C for 3 min and fixed in TAF. Samples were screened in a Leica M10 stereomicroscope, and aphelenchid nematodes were mounted on objective slides for a closer

Table 1. *Number and origin (province and municipality) of wood samples taken from pine* (Pinus sylvestris) *and spruce* (Picea abies) *and analysed for pine wood nematode* (Bursaphelenchus xylophilus) *in Norway in 2000.*

Origin Province	Municipality	Number and type of samples		
		Pine	Spruce	Total
Akershus				
	Asker	7	3	10
	Bærum	0	8	8
	Rælingen	0	1	1
Buskerud				
	Hurum	6	1	7
	Røyken	3	1	4
	Drammen	17	2	19
	Lier	11	1	12
	Ringerike	69	0	69
	Hole	55	5	60
Vestfold				
	Hof	5	17	22
Hedmark				
	Trysil	2	0	2
Total		175	39	214

examination in a Leitz DMRB interference microscope fitted with the Leica Quantimet 500MC image processing and analysis system.

Results

In this investigation nematode analyses were made of 275 wood samples. The material included 214 samples obtained from trees and wood in native forest sites, ten samples of softwood chips, 25 samples of imported lumber and 26 samples of imported solid wood packing material.

Forest wood samples

Most samples (171) were taken in the county of Buskerud and made up 80% of the total number of samples. The counties of Vestfold and Akershus had 22 and 19 samples, respectively, while the county of Hedmark was represented by two samples only (Table 1). Most samples were taken from Scots pine (*P. sylvestris*) and corresponded to 82% of the total number. With regard to pine samples, most (161) originated from Buskerud, while Akershus, Vestfold and Hedmark were represented by seven, five and two samples. The samples from Norway spruce (*P. abies*) were more evenly distributed with 17 from Vestfold, 12 from Akershus and ten from Buskerud, but none from Hedmark (Table 1).

The main fraction (153) of the samples were taken from stumps, 40 samples from detached wood, 20 from standing dead or dying trees, and one sample from roots. For pine, 153 samples (87%) were from stumps, 29 from detached wood and 19 from standing trees. For spruce, 11 samples (69%) were taken from stumps, 11 from detached wood and one sample from a standing tree.

During the survey it became apparent that the insect damage observed was caused by several species of long-horn beetles, among which were species belonging to the genera *Monochamus* and *Acanthocinus*.

Samples from wood chips

At the pulp mill of Tofte, in the municipality of Hurum in Buskerud, ten samples (Table 2) were taken from a wood chip pile. The identity of the material is unclear, but its origin could have been native or possibly mixed with imported wood.

Samples from imported material

The sampling of imported material resulted in 51 samples of material from Estonia, Russia, Spain, Lithuania, Poland, Italy, Cyprus, Israel, South Africa, Canada, USA, Argentina, Colombia, South America (unspecified), Japan and China (Table 2).

The sampling of imported lumber resulted in 25 samples. Nine samples came from spruce attacked by *Monochamus* and *Trypodendron* exported from Estonia, while seven spruce samples were taken from material imported from Russia. Two samples were from Siberian larch. In total four samples were from hardwood, with two samples from beech exported from Lithuania, one sample from oak exported from Poland and one sample from maple exported from USA. From the USA three samples were taken from unidentified conifer wood (Table 2).

Table 2. *Wood samples (n = 61) of domestic chips and imported wood analysed for pine wood nematode* (Bursaphelenchus xylophilus) *in connection with the Norwegian pine wood nematode survey in 2000. Samples are distributed on kinds of material and on exporting country.*

Country of origin	Material					
	Wood chips	Lumber				Packing
		Conifer	Spruce	Larch	Hardwood	
Argentina	..	..	..	..	..	1
Canada	..	..	..	..	..	4
China	..	..	..	..	..	1
Colombia	..	..	..	..	..	1
Cyprus	..	..	..	..	..	1
Estonia	..	..	9	..	..	
Israel	..	..	..	..	..	1
Italy	..	..	..	..	..	1
Japan	..	..	..	..	..	1
Lithuania	..	..	..	..	2	..
Poland	..	..	..	..	1	..
Russia	..	..	7	2	..	..
South Africa	..	..	..	..	..	4
South America	..	..	..	..	..	1
Spain	..	..	..	..	..	7
USA	..	3	..	..	..	..
Domestic/mixed	10	..	..	..	..	..
Total	10	3	16	2	4	26

The remaining 26 samples of import material were all taken from solid wood packing material such as wooden cases, pallets, wooden spacers and particle-boards. In most cases the actual tree species was not determined, but one pallet from USA was identified as pine. Most samples (Table 2) were taken in consignments from Spain (seven), Canada (four), South Africa (four) and USA (three). Consignments from Italy, Cyprus, Israel, Argentina, Colombia, Japan and China were represented by one sample each. One sample was taken from South American material not specified with regard to exporting country.

NEMATODE ANALYSIS OF FOREST SAMPLES

Nematodes were detected in 208 (97%) of the 214 forest samples. Aphelenchid nematodes and nematodes belonging to the order Rhabditida were the most common, but also tylenchid nematodes were observed in the material. Only six samples were found to be free from nematodes. The genus *Bursaphelenchus* was present in three (1%) of the samples.

With regard to the genus *Bursaphelenchus*, a species close to *B. sexdentati* was recorded in a wind-blown tree of *P. sylvestris* at Knutestulen, in the municipality of Drammen in Buskerud. In this case there were no signs of insect damage. An unidentified species of *Bursaphelenchus* was found in a stump of *P. sylvestris* showing signs of insect damage at Mosmoen in the municipality of Hole in Buskerud. At Svingerud in Hole, a third finding of *Bursaphelenchus* was made in a stump of *P. abies* damaged by insects.

NEMATODE ANALYSIS OF WOOD CHIPS

In the ten wood chip samples taken at the Tofte pulp mill, nematodes were detected in two samples only, and always in low numbers. The pine wood nematode, *B. xylophilus*, was not detected in this material.

NEMATODE ANALYSIS OF IMPORT MATERIAL

It was not possible to confirm the presence of *B. xylophilus* in samples taken from imported lumber, nor in samples of solid wood packing material. Nematodes of the genus *Bursaphelenchus* might, however, have been present in one sample from wood packing material of fruit from Spain. In this material there was a strong suspicion that 'dauer juveniles' of *Bursaphelenchus* were present. Due to an insufficient amount of material it was not possible to cultivate these nematodes to obtain propagative life stages for identification.

Discussion

The present survey is connected with surveys carried out in several member states of EU. These investigations are a result of the detection of PWN in forests of Portugal in May 1999 (Mota *et al.*, 1999), and the strong concern in many European countries to delimit and eradicate any infection detected in their territories.

The total sampling volume for 2000, *i.e.*, 214 samples from native forest sites, is low considering that a total number of 3000 samples has been suggested as the minimal number needed for a statistically safe statement of the presence or absence of PWN. This low sampling volume for the year 2000 needs to be compensated for by an increased survey activity in the future.

For obvious reasons the sampling was concentrated in the county of Buskerud. A large area of this county is overlapped by the two zone sites, A and B, and hence heavily exposed to a hypothetical spread of pathogens anticipated to be present in timber import material. Pine wood is an excellent substrate for the multiplication of PWN and, consequently, this tree species was the primary sampling object. All this, in combination with the high frequency of insect damage in the sampled material, demonstrates that the sampling was in compliance with the survey strategy agreed upon between Nordic scientists.

The nematode analysis demonstrated the almost ubiquitous presence of nematodes in forest wood. The occurrence of aphelenchid and rhabditid nematodes in deteriorating trees and detached wood is a normal condition. The frequent association of these nematodes with various wood-inhabiting insects ensures their transportation to and from this type of substrate. These nematodes are suspected to have a key function in the normal decomposition process.

Nematodes in the genus *Bursaphelenchus* were detected in comparatively few samples. One sample from a wind-blown tree of *P. sylvestris* contained a species possibly identical to *B. sexdentati*. This species has often been observed in pine, and has been reported from *P. sylvestris* in Austria and Germany, from *P. brutia*, *P. halepensis*, *P. nigra*, *P. pinaster* and *P. radiata* in Greece, and from *P. halepensis*, *P. pinaster* and *P. pinea* in Italy (Braasch *et al.*, 2000). The two additional observations of *Bursaphelenchus*, one of which may prove to be a new species, add to the picture of *Bursaphelenchus* spp. as a natural component of the boreal coniferous forest ecosystem. The pine wood nematode, *Bursaphelenchus xylophilus*, was not detected in any of the forest wood samples analysed.

This is in line with results from a previous survey in Norway (McNamara & Støen, 1988). In that survey nematodes were recorded in 39% of the total 429 samples. Six percent of the samples contained nematodes belonging to the genus *Bursaphelenchus*, but *B. xylophilus* was not detected.

The absence of *B. mucronatus* in the present survey was unexpected. In the previous survey (McNamara & Støen, 1988) this species was detected in one sample. That survey was not directed specifically towards material attacked by wood boring insects. The number of samples in the present survey is only half the quantity analysed by McNamara and Støen (1988), but since the proportion of samples with insect damage was high, the absence of *B. mucronatus* is still surprising, especially compared to the situation in our neighbouring countries.

In Sweden (Schroeder & Magnusson, 1989, 1992) *B. mucronatus* occurs from the province of Småland in the south to the province of Lappland in the north. Swedish data demonstrates that the area of Särna and Idre, in the province of Dalarna, has a high population of *B. mucronatus*, with 41% of the pine sawyers recorded to carry dauer juveniles (Schroeder & Magnusson, 1989). Also in Finland (Tomminen *et al.*, 1989; Tomminen, 1990) *B. mucronatus* occurs throughout the whole country. McNamara and Støen (1988) reported *B. mucronatus* from Hanestad in the county of Hedmark. The region of Särna and Idre is situated on the same latitude close to the Norwegian border, which suggests that the Hanestad nematodes were sampled from a continuous population of *B. mucronatus* in this area. In the present survey only two samples were collected from *P. sylvestris* in Hedmark. Although these trees showed signs of *Monochamus* activity, *B. mucronatus* was not detected. Extended sampling in this area would probably result in new detection of *B. mucronatus*. The data obtained so far indicate that *B. mucronatus* may have a restricted distribution in Norway.

The low frequency and densities of nematodes recorded in the samples taken from the wood chip pile at Tofte pulp mill reflects the good quality of the source material. Wood chips processed from salvage wood often have a high level of infestation.

PWN was not detected in wood chips or in the other types of import wood. In spite of this, imported coniferous wood and packing material are well-known high-risk commodities that need to be inspected and controlled more carefully. Solid wood packing material is highly dangerous and still completely out of control from the point of manufacture to the point of end-use. The presence of suspected *Bursaphelenchus* dauer juveniles in wooden boxes for fruit exported from Spain suggests that solid wood packing material, as well as coniferous wood in general exported from the Iberian Peninsula, needs to be subjected to more frequent routine inspection.

There are a number of general concerns about PWN from this survey. The total forested area of EU and Norway amounts to 146 million ha. The Nordic forests occupy almost 63 million ha, corresponding to 43% of this area. The Nordic coniferous forest area is about 49 million ha, with pine forests alone covering about 30 million ha. This represents an enormous value to the Nordic economies. The annual softwood felling in the Nordic countries alone is in the order of 40 million m^3, representing a monetary value of about US$3000 million. The Nordic forests are of fundamental importance not only to the Nordic economies, but also as a major source of wood products for the entire EU.

In half of the forest blocks visited during this survey there are trees or other types of wood with dimensions suitable for *Monochamus* oviposition, so it can be concluded that there is a good general supply of breeding material both for the insects and for the PWN. Hence there would be no restriction on nematode establishment or multiplication in Norwegian forests.

In Sweden (Magnusson & Schroeder, 1989; Schroeder & Magnusson, 1992) and Finland (Tomminen *et al.*, 1989; Tomminen, 1990), the association in nature between *Monochamus sutor* and *M. galloprovincialis*, and the native wood nematode *B. mucronatus* has been demonstrated repeatedly. Such an association is also expected to exist in Norway (McNamara & Støen, 1988). The successful transmission of *B. mucronatus* by *M. sutor* to branches and bolts of *P. sylvestris* and *P. abies* has been recorded in the laboratory (Schroeder & Magnusson, 1992). The biological similarity of *B. mucronatus* and PWN, and the fact that in Japan both species share the same vector insect *M. alternatus* (Mamiya & Enda, 1979), suggests that the natural transmission system for PWN is in full operation throughout the Nordic forest land, and most probably throughout the entire Palaearctic region.

The pine wood nematode, *B. xylophilus*, may pose a definite threat to the sustainability of the forests in vast areas of Europe. An establishment of PWN in Norway would severely affect the trade in wood products. The immediate threat to forest productivity is less well understood, but should not be ignored. Because of this there is an urgent need to detect a possible occurrence of this nematode and to eradicate such an infestation as soon as possible. This is of special importance to Norway due to the obvious difficulties in controlling larger infestations in an often rough and mountainous terrain.

Acknowledgements

The Norwegian Agricultural Inspection Service funded this survey, and this support is greatly acknowledged. We are thankful to Torstein Kvamme of the Norwegian Forest Research Institute, Aas, for his assistance in the field work. We also thank the local Plant Inspection Services in Oslo, Kristiansand, Stavanger and Trondheim, for help in providing samples of imported wood materials. The laboratory technicians, Irene Rasmussen and Kari Ann Strandenæs, at the Nematode Laboratory of the Norwegian Crop Research Institute, are greatly acknowledged for their efficient handling of samples and extractions.

References

ANON. (2000). E.C. Pinewood Nematode Survey Protocol 2000. *European Commission, Directorate-General Health and Consumer Protection, Directorate E – Public, animal and plant health. Unit E1. Legislation relating to crop products and animal nutrition. SANCO E/1 D(00)*, 7 pp.

BRAASCH, H., CAROPPO, S., SKARMOUTSOS, E. & TOMICZEK, C. (2000). Occurrence, host range, distribution and vectors of *Bursaphelenchus* species in coniferous trees in Europe. In: *Pest risk analysis of pinewood nematode related* Bursaphelenchus *species in view of south European pine wilting and wood imports from Asia.* FAIR CT 95-0083, 68 pp.

EVANS, H.F., MCNAMARA, D.G., BRAASCH, H., CHADOEUF, J. & MAGNUSSON, C. (1996). Pest Risk Analysis (PRA) for the territories of the European Union (as PRA area) on *Bursaphelenchus xylophilus* and its vectors in the genus *Monochamus. EPPO Bulletin* 26, 199-249.

MAGNUSSON, C. & SCHROEDER, L.M. (1989). First record of a *Bursaphelenchus* species (Nematoda) in *Monochamus* beetles in Scandinavia. *Anzeiger für Schädlingskunde, Pflanzenschutz und Umweltschutz* 62, 53-54.

MAMIYA, Y. & ENDA, N. (1979). *Bursaphelenchus mucronatus* n. sp. (Nematoda: Aphelenchoididae) from pinewood and its biology and pathogenicity to pine trees. *Nematologica* 25, 353-361.

MCNAMARA, D.G. & STØEN, M. (1988). A survey for *Bursaphelenchus* spp. in pine forests in Norway. *EPPO Bulletin* 18, 353-363.

MOTA, M.M., BRAASCH, H., BRAVO, M.A., PENAS, A.C., BURGERMEISTER, W., METGE, K. & SOUSA, E. (1999). First report of *Bursaphelenchus xylophilus* in Portugal and Europe. *Nematology* 1, 727-734.

SCHROEDER, M. & MAGNUSSON, C. (1989). Tallvednematoden – ett hot mot svensk skog? *Skogsfakta, Biologi och skogsskötsel 64.* SLU Uppsala, Sweden, 4 pp.

SCHROEDER, L.M. & MAGNUSSON, C. (1992). Transmission of *Bursaphelenchus mucronatus* (Nematoda) to branches and bolts of *Pinus sylvestris* and *Picea abies* by the cerambycid beetle *Monochamus sutor*. *Scandinavian Journal of Forest Research* 7, 107-112.

TOMMINEN, J. (1990). Presence of *Bursaphelenchus mucronatus* (Nematoda: Aphelenchoididae) fourth dispersal stages in selected conifer beetles in Finland. *Silva Fennica* 24, 273-278.

TOMMINEN, J., NUORTEVA, M., PULKKINEN, M. & VÄKEVÄ, J. (1989). Occurrence of the nematode *Bursaphelenchus mucronatus* Mamiya & Enda, 1979 (Nematoda: Aphelenchoididae) in Finland. *Silva Fennica* 23, 271-277.

Nematology Monographs & Perspectives, 2003, Vol. 1, 113-126

Surveying and recording of nematodes of the genus *Bursaphelenchus* in conifer forests in Greece and pathogenicity of the most important species

Helen MICHALOPOULOS-SKARMOUTSOS,
George SKARMOUTSOS,
Maria KALAPANIDA and A. KARAGEORGOS

Forest Research Institute, Agricultural Research Foundation, 57006 Thessaloniki, Greece

Summary – Between 1996 and 2001, surveys were carried out in Greece in order to record and determine the *Bursaphelenchus* fauna on conifers. This was done in the framework of three projects, namely: *i*) FAIR 1CT 95-0083; *ii*) Directive 2000 of the EU; and *iii*) a bilateral project with Bulgaria. Altogether 364 wood samples from all over Greece were investigated for the presence of *Bursaphelenchus* nematodes. The pine wood nematode *B. xylophilus* was not found. The species isolated were: *Bursaphelenchus leoni, B. sexdentati, B. teratospicularis, B. eggersi, B. hellenicus* and *B. mucronatus*. These were the first *Bursaphelenchus* surveys in Greece and all species constitute new records in this country. Attempts to rear the above nematodes in the laboratory have proved successful for *B. leoni, B. sexdentati* and *B. hellenicus*. This was done in Petri dishes on *Botrytis cinerea* cultures for further checking of their pathogenicity by inoculation tests on 3-year-old seedlings of *P. brutia, P. halepensis, P. pinaster* and *P. sylvestris*. The results showed that *B. sexdentati* is highly virulent, causing mortality of up to 100% of the inoculated seedlings, followed by *B. leoni*, whereas *B. hellenicus* is considered as non-pathogenic.

Wilting of different species of pine, of unclear aetiology, has been observed in Greece during the last 20 years. The first ever record of the isolation of nematodes of the genus *Bursaphelenchus* from *Pinus brutia* and *P. nigra* was made in 1987. However, at that time it was not possible to identify the nematodes to species level (Skarmoutsos & Michalopoulou, 1987). Initially fragmentary work has become systematic only during the last 5 years. The recording of the nematode fauna of conifer species was effected in the framework of three different research projects, namely: *i*) FAIR 1CT 95-0083 'Pest Risk Analysis of pine wood nematode related *Bursaphelenchus* species in view of South European pine wilting and wood imports from Asia', funded by the EU; *ii*) Directive 2000 of the EU; and *iii*) a bilateral project between Greece and Bulgaria in the framework of Cooperation on Research and Development between the two countries.

Efforts were undertaken to rear the nematodes isolated in pure cultures in order to further check their pathogenicity by inoculation tests on 3-year-old pine seedlings. In this paper the results of the surveys and those of the inoculation tests are presented.

Materials and methods

NEMATODE SURVEY

In order to determine the *Bursaphelenchus* fauna of conifers in Greece, surveys were carried out between 1996 and 2001. Emphasis was given to areas with symptoms of pine decline and to those neighbouring ports where timber is imported or wood processing factories exist. From such areas sampling of apparently healthy trees also took place. From areas with wilting symptoms, samples were taken from dying or recently dead trees, *i.e.*, for not more than 6 months before sampling. Sampling was preferably done at parts of the trees showing damage from insect attacks or affected by blue stain. Samples were taken from three different heights along the trunk of the tree, *i.e.*, from the lower, medium and upper parts and consisted of disks of a thickness of *ca* 3 cm each. In cases where trees could not be felled, samples were taken with the use of a Pressler's borer. Sampling at each area involved from one up to ten trees, the decision being made according to the number of affected trees in the area. Individual samples from each area were combined to form a mixed sample. Samples were put into plastic bags and were transferred to the laboratory. At the laboratory the disks were cut into small pieces and were submitted to extraction for nematodes or were kept in dark conditions in a cooling chamber at 4°C until processed.

Conifer species that were examined during the three projects above included: *Pinus nigra, P. brutia, P. halepensis, P. sylvestris, P. pinaster, P. pinea, P. radiata, Abies cephalonica, A. borisii regis* (= *A. hybrido-*

genus, a natural hybrid of *A. alba* × *A. cephalonica*) and *Cupressus sempervirens*. At the same time sawdust of different pine species from wood processing factories, as well as conifer timber imported from abroad, were also examined. All three projects were the responsibility of the Forest Pathology Laboratory of the Forest Research Institute in Thessaloniki, Greece, while the contribution of the Forest Service in collecting the samples has been important.

Extraction and identification

For extraction purposes the wood disks collected were further cut into thinner disks by electric saw and then by hand into smaller pieces of 5-10 mm in size with the use of pruning scissors. About 30 g of fresh wood per tree were put to extraction. In the case of mixed samples, the actual weight for extraction purposes consisted of 30 g multiplied by the number of trees which constituted the mixed sample. Extraction was carried out with the modified Baermann funnel technique for 48 h at room temperature at 20-25°C. At the end of this period, 10 ml of the potential nematode suspension was collected from the base of each funnel. Checking of the suspension for presence of nematodes was first done under a stereo microscope. In most of the cases, apart from *Bursaphelenchus* spp., other nematode genera that live in the wood were also present. The *Bursaphelenchus* species were selected for further examination with a light microscope under high magnification. The identification was documented by drawings, microphotographs, permanent slides and by molecular biological methods which were undertaken at BBA at Braunschweig, Germany. In cases where *Bursaphelenchus* species were found in sufficient numbers, further efforts were undertaken to rear them under laboratory conditions (Braasch *et al.*, 2000).

Nematode culturing

Rearing of the nematodes was done on cultures of *Botrytis cinerea* in 9 cm Petri dishes. Both the sporulating and non-sporulating forms were used. The substrate for growing the fungus was either 2% Malt Extract Agar (MEA) or 3.9% Potato Dextrose Agar (PDA). From 40 to 50 nematodes of each species that had undergone several washings in distilled and sterilised water were transferred to the fungus culture and covered by a sterile microscope cover slip (Southey, 1986). The lids of the Petri dishes were then closed with parafilm. Subsequent subcultures were made by transferring pieces of old cultures to new Petri dishes with freshly grown colonies of *B. cinerea* (Braasch *et al.*, 2000).

Inoculation of plants

The *Bursaphelenchus* species that could be reared in the laboratory were *B. sexdentati, B. leoni* and *B. hellenicus*. Their pathogenicity towards different conifer species was checked during the years 1997-1999. Inoculation tests were carried out both in climate chambers (temperature 25°C, RH 60%) and outdoors. Conifer species used included *P. sylvestris, P. nigra, P. brutia. P. pinaster* and *P. halepensis*. The 3-year-old seedlings of the above pine species were transplanted into 10 × 10 cm plastic pots 2 months before inoculation, and were placed in climate chambers or outdoors. Outdoors the transplanted seedlings were caged under fine mesh. The plants in the climate chambers were watered with 100 ml of water twice weekly. Outdoor plants were irrigated according to their needs. Twenty plants were inoculated with each *Bursaphelenchus* isolate, while ten similar plants were used as control and received only distilled water (Braasch *et al.*, 2000). Inoculations were carried out as described by Braasch (1996).

The nematodes that were multiplied as described above were obtained from the Petri dishes by washing with a stream of distilled and sterilised water on the day of inoculation. About 6000 nematodes in 0.5 ml of water were inoculated on each plant. The success of the inoculation was evaluated by weekly assessment of the symptoms and reisolation of the nematodes after the plants had died or after 3 months at the latest. To this effect the plants were cut at 2 cm above ground and the stems were submitted to extraction (Braasch *et al.*, 2000).

In order to compare the results and evaluate the host suitability of the 3-year-old seedlings that had been inoculated, the modified index of Relative Host Suitability (RHS) suggested by Braasch (1977) was used. In this case RHS was the product of the percentage of successfully inoculated plants (plants from which nematodes were re-isolated) and the average number of reisolated nematodes per g of fresh wood of the stems of the inoculated plants, divided by 1000.

Results

BURSAPHELENCHUS SPECIES FOUND IN GREECE

In the framework of the three projects referred to at the beginning, a total of 365 wood samples were examined from declining, apparently healthy trees, sawdust and imported conifer wood from all over Greece. From the investigated samples, 104 (28%) contained different *Bursaphelenchus* species. Altogether six species were found, *i.e.*, *B. leoni, B. sexdentati, B. teratospicularis, B. hellenicus, B. mucronatus* and *B. eggersi*. To the above six species must be added *B. tusciae* which was recently isolated for the first time in Greece. However, from nine samples it has not been possible to identify nematodes to the species level due to the small numbers of specimens and their poor condition under the microscope.

Fig. 1 shows the localities where *Bursaphelenchus* species were found in Greece. Table 1 shows the forest species examined, the number of samples in each health category, the number and the percentage of samples from which *Bursaphelenchus* species were either isolated or not, and the nematode species isolated from each host. Table 2 shows the number and frequency of *Bursaphelenchus* species which were found in the wood samples.

The most frequently found species was *B. sexdentati*, followed by *B. leoni. Bursaphelenchus xylophilus* was in no case isolated, while in a great number of samples more than one species of *Bursaphelenchus* were present.

REARING OF EXTRACTED *BURSAPHELENCHUS* SPECIES

Efforts were undertaken in the laboratory to rear the *Bursaphelenchus* species isolated. Successful multiplication was recorded for *B. sexdentati, B. leoni* and *B. hellenicus* on *B. cinerea*, while it was not possible to multiply *B. teratospicularis*. At the same time, *B. eggersi* and *B. mucronatus* were represented by small numbers of specimens and therefore such efforts were not made.

PATHOGENICITY TESTS

Tables 3 and 4 show the results of the inoculation tests both in the climate chamber and outdoors. Pathogenicity tests with *B. sexdentati*, the

Table 1. *Incidence of* Bursaphelenchus *species in samples from tree species in Greece.*

Tree species	condition	No of samples	*Bursaphelenchus* positive		Species isolated
			n	%	
Pinus brutia	wilting	88	33	37.5	*B. eggersi, B. hellenicus, B. leoni, B. mucronatus, B. sexdentati, B. teratospicularis, B.* sp.
	healthy	34	0	0	
	total	122	33	27	
Pinus nigra	wilting	64	31	48	*B. leoni, B. sexdentati, B.* sp.
	healthy	29	0	0	
	total	93	31	33	
Pinus pinaster	wilting	30	17	57	*B. eggersi, B. leoni, B. sexdentati, B. teratospicularis, B.* spp.
	healthy	3	0	0	
	total	33	17	52	
Pinus sylvestris	wilting	27	0	0	–
	healthy	4	0	0	
	total	31	0	0	
Pinus halepensis	wilting	35	20	57	*B. hellenicus, B. leoni, B. sexdentati, B. teratospicularis*
	healthy	9	0	0	
	total	44	20	45	
Pinus radiata	wilting	7	3	43	*B. leoni, B. sexdentati*
	healthy	1	0	0	
	total	8	3	57.5	
Pinus pinea	wilting	0	–	–	–
	healthy	2	0	0	
	total	2	0	0	
Abies cephalonica	wilting	6	0	0	–
	healthy	0	–	–	
	total	6	0	0	
Abies borisii regis	wilting	4	0	0	–
	healthy	6	0	0	
	total	10	0	0	

Table 1. *(Continued).*

Tree species condition		No of samples	*Bursaphelenchus* positive		Species isolated
			n	%	
Cupressus semper-virens	wilting	1	0	0	–
	healthy	0	–	–	
	total	1	0	0	
Pinus spp.	sawing dust	11	0	0	–
	imported wood	4	0	0	–
Total		365	104	28.0	

Table 2. *Analysis of* Bursaphelenchus *species (Bl:* B. leoni*; Bh:* B. hellenicus*; Bs:* B. sexdentati*; Bt:* B. teratospicularis*; Bm:* B. mucronatus*; Be:* B. eggersi*; Bspp: unidentified) found in relation to host tree species.*

Samples with *Bursaphelenchus*		*Bursaphelenchus* species													
		Bl		Bh		Bs		Bt		Bm		Be		Bspp	
Host	n	n	%	n	%	n	%	n	%	n	%	n	%	n	%
Pinus brutia	33	13	39	5	15	24	72	5	15	1	3	2	6	1	3
P. nigra	31	17	55	–	–	20	65	–	–	–	–	–	–	6	19
P. pinaster	17	8	47	–	–	11	65	1	6	–	–	1	6	2	12
P. sylvestris	–	–	–	–	–	–	–	–	–	–	–	–	–	–	–
P. halepensis	20	10	50	2	10	15	80	2	10	–	–	–	–	–	–
P. radiata	3	3	100	–	–	3	100	–	–	–	–	–	–	–	–
Total	104	51	49	7	7	74	71	8	8	1	1	3	3	9	9

most frequently occurring species, provoked a high degree of mortality, reaching 100% on *P. pinaster*, 90-100% on *P. nigra* and 100% on *P. sylvestris*. *Bursaphelenchus hellenicus*, on the other hand, appears to be non-pathogenic on *P. brutia*, *P. pinaster* and *P. sylvestris* on which it was tested. Finally *B. leoni* tested outdoors on *P. halepensis* provoked a mortality of 55%. The progress of the infections in which mortality was caused is shown in Fig. 2.

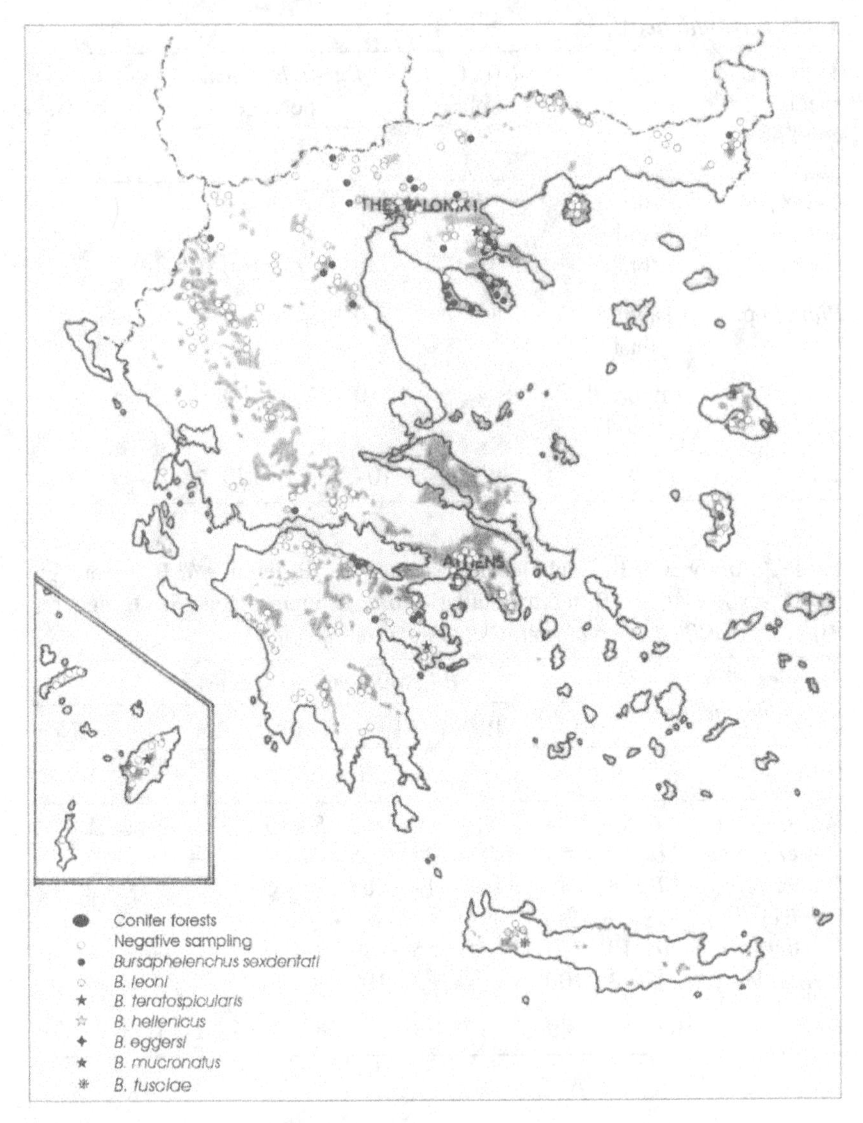

Fig. 1. *Occurrence of* Bursaphelenchus *species in Greece.*

Discussion

The surveys on the occurrence of *Bursaphelenchus* species in Greece contribute to the knowledge on the distribution of this genus in Europe and were the first ever undertaken in this country. Six species of

Bursaphelenchus were recorded, namely: *B. sexdentati, B. leoni, B. teratospicularis, B. mucronatus, B. hellenicus* and *B. eggersi*, which are all a first record in Greece, while among them *B. hellenicus* is a species described for the first time in the framework of FAIR 1CT 95-0083 (Skarmoutsos *et al.*, 1998).

The species found in Greece appear to be typical representatives of the Mediterranean *Bursaphelenchus* fauna (Philis & Braasch, 1996; Braasch 2000). The most frequently found species were *B. sexdentati* and *B. leoni*, while *B. mucronatus*, which appears frequently in central and northern European countries, has been isolated only once in Greece and in very small numbers.

All isolations of *Bursaphelenchus* species were made from different pine species and in no case from other conifers such as *Abies* or *Cupressus* that were also examined. Pine species which were found to contain *Bursaphelenchus* spp. were *P. brutia, P. nigra, P. pinaster, P. halepensis* and *P. radiata*, while *P. sylvestris* and *P. pinea* did not yield any nematodes. All nematodes were isolated from dead or dying pines with wilting symptoms, insect attack mainly by scolytids, and blue stain of the wood. Nematodes were not isolated from healthy trees or from samples of sawdust or from timber from wood processing factories.

From the *Bursaphelenchus* species isolated, those of *B. sexdentati* and *B. hellenicus* were easily multiplied in pure cultures in the laboratory, while *B. leoni* did so also but with some difficulties. *Bursaphelenchus teratospicularis* could not be reared under laboratory conditions although sufficient numbers of specimens were available.

The results of the inoculation tests show that *B. sexdentati* (four local isolates) is a strong pathogen on *P. pinaster, P. nigra* and *P. sylvestris* seedlings both in the climate chamber and outdoors. At the same time RHS values were higher in outdoor tests, probably because outdoor temperatures during summer were higher than those in the climate chambers. *Bursaphelenchus hellenicus* appears to be non-pathogenic on *P. brutia, P. pinaster* and *P. sylvestris* seedlings either in the climate chamber and outdoors. Finally, *B. leoni* shows a relatively low virulence on *P. halepensis* seedlings on which it was tested.

From the above it can be concluded that *B. sexdentati* may be the agent responsible or one of the factors that contribute to pine decline in Greece. However, such decline phenomena are of a small scale and not widespread, while their intensity differs over the course of successive years. It is thought that the pine decline phenomena in this country,

Table 3. *Inoculation tests of three tree species* (Pinus pinaster *(Pp),* P. nigra *(Pn) and* P. brutia *(Pb)) with* Bursaphelenchus sexdentati *(GR-5(W), GR-7(W) and GR-8(W)) and* B. hellenicus *(GR-2(W)) in a climatic chamber (25°C).*

Experiment date	Nematode	Plants inoculated						Nematodes reisolated from successfully inoculated plants		RHS
		Species	Total (n)	Mortality		Success		per plant	per g wood	
				(n)	(%)	(n)	(%)			
03.7.97	GR-5(W)	Pp	20	20	100	20	100	970	191	19
	control		10	1	10			0		
22.8.97	GR-7(W)	Pn	20	18	90	20	100	1102	336	34
	control		10	0	0			0		
22.8.97	GR-8(W)	Pn	20	18	90	18	90	1227	317	29
	control		10	0	0			0		
07.7.98	GR-2(W)	Pb	20	0	0	8	40	2	1	0
	control		10	0	0			0		
07.7.98	GR-2(W)	Pp	20	0	0	0	0	0	0	0
	control		10	0	0			0		

RHS = (% successfully inoculated plants × number of nematodes reisolated per g of fresh weight)/1000.

Table 4. *Inoculation tests of five tree species* (Pinus pinaster *(Pp)*, P. nigra *(Pn)*, P. silvestris *(Ps)*, P. brutia *(Pb) and* P. halepensis *(Ph)) with* Bursaphelenchus sexdentati *(GR-5(W), GR-7(W), GR-8(W) and G-31(W))*, B. hellenicus *(GR-2(W)) and* B. leoni *(G-42(W)) outdoors.*

Experiment date	Nematode	Plants inoculated						Nematodes reisolated from successfully inoculated plants		RHS
		Species	Total (n)	Mortality		Success		per plant	per g wood	
				(n)	(%)	(n)	(%)			
11.7.97	GR-5(W)	Pp	20	20	100	20	100	27 200	1230	123
	control		10	1	10			0		
22.8.97	GR-7(W)	Pn	20	20	100	20	100	1820	633	63
	control		10	0	0			0		
22.8.97	GR-8(W)	Pn	20	20	100	20	100	2760	876	88
	control		10	0	0			0		
28.7.98	G-31(W)	Ps	20	20	100	20	100	11 282	3137	314
	control		10	0	0			0		
07.7.97	GR-2(W)	Pb	20	2	10	17	85	9	4	0
	control		10	0	0			0		
07.7.98	GR-2(W)	Pp	20	7	35	4	20	5526	320	1
	control		10	2	20			0		
07.7.97	GR-2(W)	Ps	20	0	0	0	0	0	0	0
	control		10	0	0			0		
15.6.99	GR-2(W)	Ps	20	0	0	0	0	0	0	0
	control		10	0	0			0		
13.8.98	G-42(W)	Ph	20	11	55	11	55	264	204	6
	control		10	1	10			0		

RHS = (% successfully inoculated plants × number of nematodes reisolated per g of fresh weight)/1000.

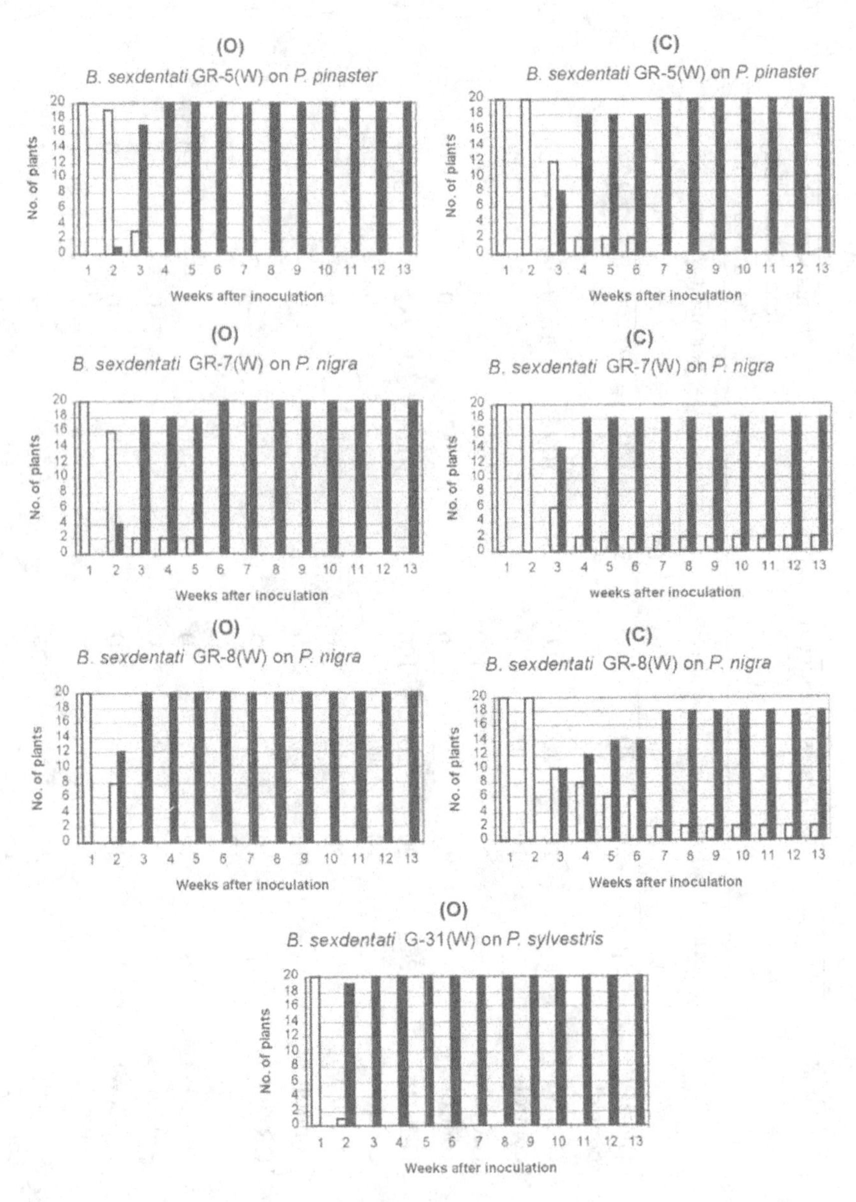

Fig. 2. *Progress of infection in five* Pinus *species inoculated with two* Bursaphelenchus *species and grown outdoors (O) or at 25°C in climate chambers (C).*

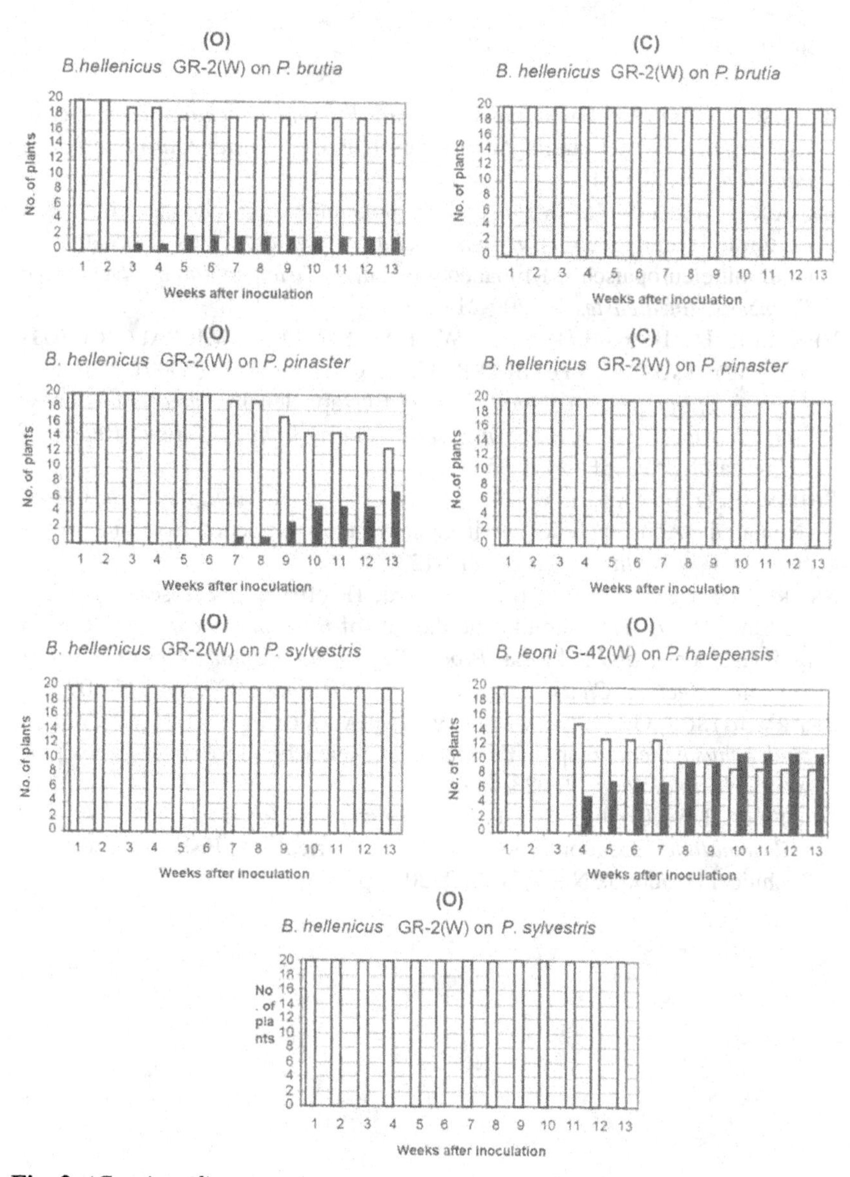

Fig. 2. *(Continued).*

which undoubtedly need further research, are of a complex nature in which insects, fungi and nematodes are involved, while soil and climatic conditions also play a significant role.

References

BRAASCH, H. (1996). Pathogenitätstests mit *Bursaphelenchus mucronatus* an Kiefern- und Fichtensämlingen in Deutschland. *European Journal of Forest Pathology* 26, 205-216.

BRAASCH, H. (1997). Wirts und Pathogenitätuntersuchungen mit dem Kiefernholznematoden (*Bursaphelenchus xylophilus*) aus Nordamerika unter mitteleuropäischen Klimabedigungen. *Nachrichtenblatt des Deutschen Pflanzenschutzdienstes* 49, 209-214.

BRAASCH, H., BURGERMEISTER, W., HARMEY, M.A., MICHALOPOULOS-SKARMOUTSOS, H., TOMICZEK, C. & CAROPPO, S. (2000). *Pest Risk Analysis of pinewood nematode related* Bursaphelenchus *species in view of South European pine wilting and wood imports from Asia.* Final Report of EU research project Fair CT 95-0083.

PHILIS, J. & BRAASCH, H. (1996). Occurrence of *Bursaphelenchus leoni* (Nematoda, Aphelenchoididae) in Cyprus and its extraction from pine wood. *Nematologia Mediterranea* 24, 119-123.

SKARMOUTSOS, G. & MICHALOPOULOU, H. (1987). [Necroses of pines at Kedrinos Hill of Thessaloniki and the role of *Bursaphelenchus* sp.] '*Forests of* P. halepensis *and* P. brutia' *Proceedings of the Hellenic Forestry Society Scientific Meeting, Chalkis, 30 September-2 October 1987*, pp. 297-309.

SKARMOUTSOS, G., BRAASCH, H. & MICHALOPOULOU, H. (1998). *Bursaphelenchus hellenicus* sp. n. (Nematoda Aphelenchoididae) from Greek pine wood. *Nematologica* 44, 623-629.

SOUTHEY, J.F. (ED.) (1986). *Laboratory methods for work with plant and soil nematodes.* London, UK, Ministry of Agriculture, Fisheries and Food Technical Handbook No. 7, HMSO, 202 pp.

Nematology Monographs & Perspectives, 2003, Vol. 1, 127-143

Morphology of *Bursaphelenchus xylophilus* compared with other *Bursaphelenchus* species

Helen BRAASCH

Department for National and International Plant Health, Federal Biological Research Centre for Agriculture and Forestry, Kleinmachnow Branch, Stahnsdorfer Damm 81, 14532 Kleinmachnow, Germany

Summary – The genus *Bursaphelenchus* contains about 55 species, approximately three quarters of them living in conifer trees throughout the Northern Hemisphere. The member states of the European Union have begun a survey of forest nematodes in order to try to ensure early detection of any further establishment of the pine wood nematode in Europe. Other *Bursaphelenchus* spp. are being, and will continue to be, found in this survey and, therefore, it is important to be able to clearly distinguish them from *B. xylophilus*. The *Bursaphelenchus* fauna of Europe shares many species with that of Asia, but is apparently clearly different from that of North America. Morphological studies of *Bursaphelenchus* spp. revealed that several species groups can be distinguished among the 28 species detected in Europe so far. The number of incisures in the lateral field, spicule shape, number and position of caudal papillae, presence and size of a vulval flap and the shape of female tails are essential diagnostic features. The members of the *B. xylophilus* group (*B. xylophilus*, *B. mucronatus*, *B. fraudulentus*) can easily be separated from all other species by the presence of four lateral lines, the typical shape of spicules, the special position of the caudal papillae and the large vulval flap of females. Within this group, species can be morphologically distinguished by the shape of the female tail end. The mucronate form of *B. xylophilus*, which is morphologically similar to *B. mucronatus*, does not occur in Europe. *Bursaphelenchus* species showing the typical features of the *B. xylophilus* group and having a round female tail end can, therefore, be definitely morphologically identified as *B. xylophilus*.

The pine wood nematode, *Bursaphelenchus xylophilus* (Steiner & Buhrer, 1934) Nickle, 1970, is the causal agent of pine wilt disease, but can also persist in living trees. Other *Bursaphelenchus* species vectored by various wood and bark breeding beetles may be found in insect-attacked living or dead trees. Approximately 75% of the 55 species known worldwide can be found in coniferous trees. Twenty-eight species are known to occur in conifers in Europe (Braasch, 2001). The pan-European monitoring survey for *B. xylophilus*, which was started after the detection of the pine wood nematode in Portugal (Mota *et al.*, 1999), makes the reliable differentiation of this species from related species indispensable. Whereas the morphology of *B. xylophilus* is well known (Nickle *et al.*, 1981; Yik & Birchfield, 1981), several other species are poorly described, and some of them show characters similar to the pine wood nematode. Molecular methods are available, for example, use of ITS-RFLP technique (Hoyer *et al.*, 1998), but they are not accessible to all laboratories and are more expensive than morphological methods. So far, molecular patterns have not been described for all European species.

Bursaphelenchus species are originally inhabitants of the Northern Hemisphere and most probably linked to conifers in their evolution. The *Bursaphelenchus* fauna of Europe shares many species with that of Asia, but is apparently clearly different from that of North America. The quarantine pest pine wood nematode is, however, a potential threat to coniferous forests around the world. The purpose of this paper is to assist the identification of *Bursaphelenchus* spp. found in coniferous wood in Europe.

Material and methods

Most of the *Bursaphelenchus* species included in this study were collected within the framework of an EU-supported research project ('Pest Risk Analysis of pinewood nematode-related *Bursaphelenchus* species in view of south European pine wilting and wood imports from Asia') in Austria, Germany, Greece and Italy (Braasch *et al.*, 2000) and during a monitoring survey for *B. xylophilus* executed in Germany in summer of 2000 (Schönfeld *et al.*, 2001) and studied with the aid of a Zeiss Axioskop microscope and using a Sony CCDmIRIS video camera. The microphotographs were produced by M. Brandstetter, Federal Forest Research Centre, Institute for Forest Protection, Vienna, Austria, by use of a scanning electron microscope Zeiss DSM 940 with cryopreservation Oxford G 1500. Several species were cultivated on *Botrytis cinerea*.

Specimens were fixed in TAF and preserved in glycerine. Information on other *Bursaphelenchus* species was obtained solely from the study of pertinent references.

Results

GROUPING OF *BURSAPHELENCHUS* SPP.

Morphological studies of *Bursaphelenchus* spp. by light and scanning electron microscopy, by microphotography and by comparison with literature, revealed that several species groups with morphologically-related species can be distinguished in Europe (Table 1). The number of incisures in the lateral field, clearly visible by scanning electron microscopy, is considered as a basic feature for grouping. Other essential diagnostic features are spicule shape, number and position of caudal papillae, presence and size of a vulval flap and the shape of female tails. Five species (*B. abietinus, B. fungivorus, B. idius, B. cryphali, B. teratospicularis*) either do not correspond with any other species with regard to these characters or the features of these characters are not fully known and, therefore, cannot yet be grouped together with other species. Four main groups can be distinguished represented by species with two, three, four and six lateral lines. Most of the species belong to the two main groups with three and four lateral lines.

Within each of these two main groups, several species groups can be differentiated according to the characters mentioned above. They are named *B. eggersi* group, *B. leoni* group and *B. hofmanni* group within the main group with three lateral lines, and *B. xylophilus* group, *B. sexdentati* group and *B. fungivorus* within the main group with four lateral lines. The names were selected in accordance with the frequency and, in the case of the *B. xylophilus* group, the importance of the relevant species. The members of the *B. xylophilus* group, namely *B. xylophilus, B. mucronatus* and *B. fraudulentus*, can easily be separated from all other species by the presence of four lateral lines, the typical shape of spicules (Table 2; Fig. 1), the special position of the caudal papillae (Fig. 2) and the large vulval flap of females (Figs 1, 2). They show a single preanal, a pair of adanal and a double pair of post-anal caudal papillae just before the bursa. A double pair of post-anal caudal papillae is also present in *B. fungivorus*: these papillae are, however, differently arranged to each other (longitudinally shifted) compared with *B. xylophilus*, whereas the

Table 1. *Diagnostic features of groups of European* Bursaphelenchus *species in conifers.*

Lateral lines	Species/species group characteristics
2	*B. abietinus* Caudal papillae: two pairs adanal, one pair postanal just before bursa, characteristic spicules, small vulval flap
3	*B. eggersi* group (*B. eggersi, B. tusciae, B. glochis*) Caudal papillae: one single and one pair preanal, one pair postanal in about the middle of the tail, one pair at bursa, straight spicules, small vulval flap, female tail end ventrally bent
3	*B. leoni* group (*B. leoni, B. silvestris**, *B. eidmanni**) Caudal papillae: one single (not known in *B. silvestris*) and one pair preanal, one pair postanal in about the middle of the tail, one pair at bursa, condylus of spicules dorsally bent, large vulval flap, long and thin fermale tail
3	*B. hofmanni group* (*B. hofmanni, B. hellenicus, B. sachsi**, *B. nuesslini**, *B. pinasteri, B. chitwoodi**, *B. paracorneolus*) Caudal papillae: one single (not known in all cases) preanal, one pair preanal or adanal, one pair postanal in distance to bursa, one pair at bursa, various shapes of more or less bent spicules, small vulval flap
4	*B. xylophilus* group (*B. xylophilus, B. mucronatus, B. fraudulentus*) Caudal papillae: one single preanal, one pair adanal, double pair of postanal caudal papillae just before bursa, characterisic spicules, large vulval flap
4	*B. sexdentati* group (*B. sexdentati, B. poligraphi, B. pinophilus, B. incurvus**, *B. naujaci, B. piniperdae**, *B. "borealis"*) Caudal papillae: one single preanal, one pair adanal, two pairs postanal (one before bursa, one at bursa), spicules bent and relatively compact, small vulval flap, protuberance behind vulva
4	*B. fungivorus* Caudal papillae: one single preanal, one pair adanal, double pair of postanal caudal papillae before bursa, characteristic shape of compact spicules, no vulval flap, female tail long, tapering, ventrally bent

Table 1. *(Continued).*

Lateral lines	Species/species group characteristics
6	*B. idius* Caudal papillae: one single and one pair preanal, one pair just postanal, one pair before bursa, delicate spicules, vulval flap absent
Un-known	*B. cryphali, B. teratospicularis* Caudal papillae: one pair preanal, one pair before bursa, various shape of spicules characteristic for each species, vulval flap absent, *B. teratospicularis* with short postuterine branch

* Number of lateral lines not known, attachment to the group because of other morphological features.

B. sexdentati group has no double pair. Additionally, the females of *B. fungivorus* have no vulval flap, whereas the females of the *B. sexdentati* group have one, but their flaps are not as large as in the case of the *B. xylophilus* group.

Species identification

Adults of the genus *Bursaphelenchus* can be recognised by the general characters of aphelenchid nematodes and by following clearly visible features of the genus: cephalic region relatively high and offset from the body by a constriction, 10-20 μm long stylet usually with small basal swellings, pharyngeal glands overlapping intestine, vulva at 70-80% of body length, rosethorn-shaped spicules, conoid and strongly ventrally curved male tail with a variously pointed terminus like a strong thorn, small terminal bursa of variable shape (seen only in dorso-ventral position). Provided these characters are recognised, the most easily identifiable and specific features of each species are the shape of the male spicule and female tail (Table 2). In most cases, these features vary only slightly, whereas the shape of the bursa is very variable. The features have proved suitable for a preliminary species identification and may be used directly or after determination of group affiliation. Comparison of the result with the original description of a species may be helpful. Morphometrics of *B. xylophilus* vary considerably and are given in Table 3.

Table 2. *Shape of spicules and female tail ends of European* Bursaphelenchus *species in conifers.*

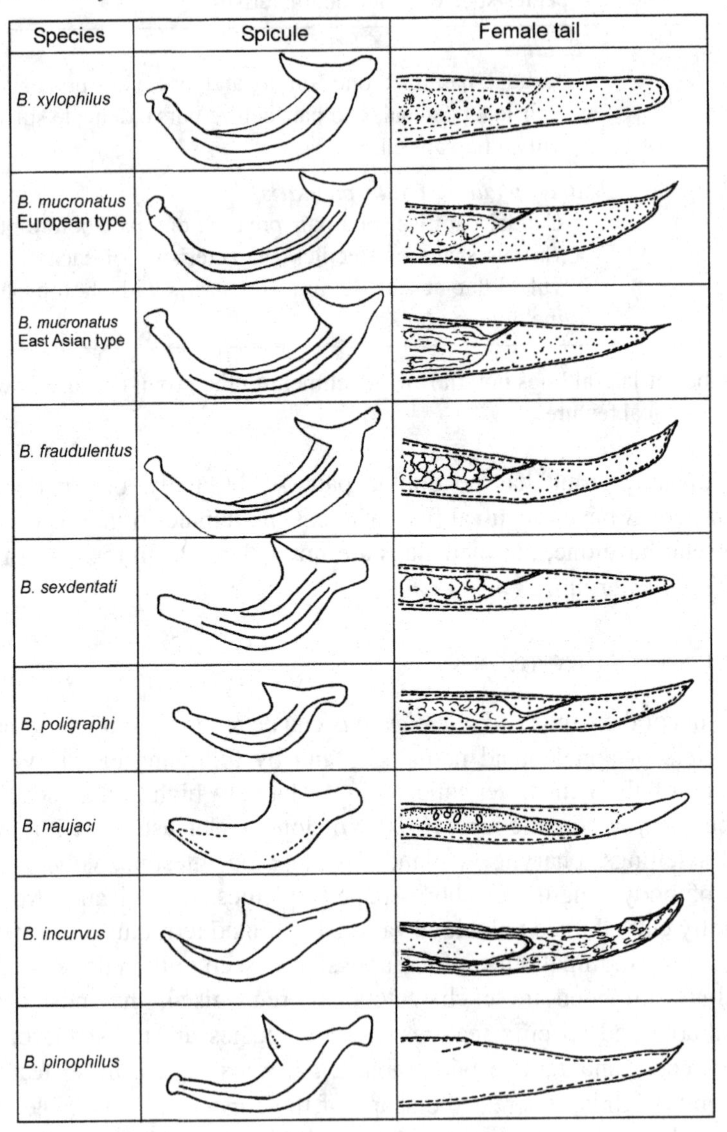

Species	Spicule	Female tail
B. xylophilus		
B. mucronatus European type		
B. mucronatus East Asian type		
B. fraudulentus		
B. sexdentati		
B. poligraphi		
B. naujaci		
B. incurvus		
B. pinophilus		

Whereas the species of the groups named after *B. eggersi, B. leoni, B. xylophilus* and *B. sexdentati* show considerable homogeneity, the members of the *B. hofmanni* group are not so homogeneous, and some

Table 2. *(Continued).*

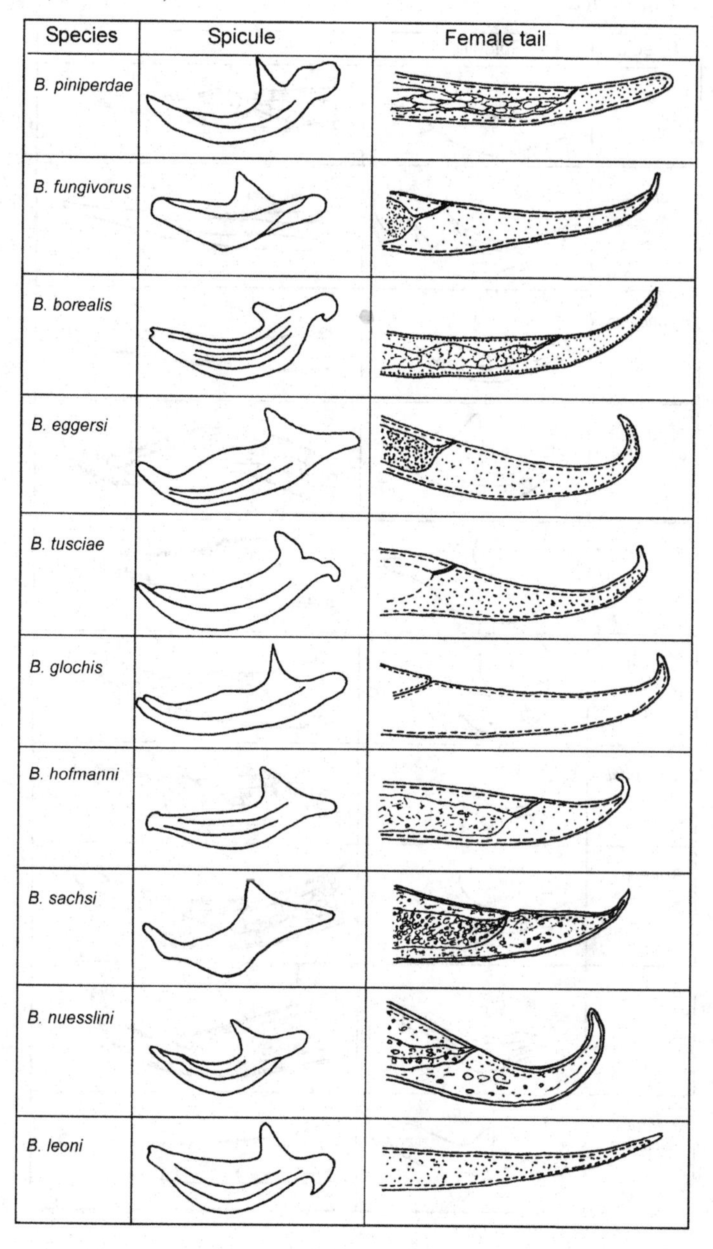

Species	Spicule	Female tail
B. piniperdae		
B. fungivorus		
B. borealis		
B. eggersi		
B. tusciae		
B. glochis		
B. hofmanni		
B. sachsi		
B. nuesslini		
B. leoni		

Table 2. *(Continued).*

Species	Spicule	Female tail
B. eidmanni	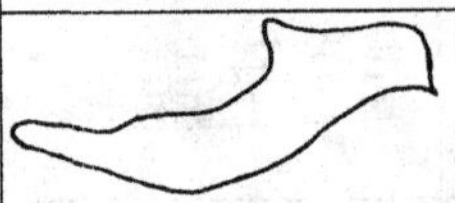	
B. silvestris		
B. hellenicus		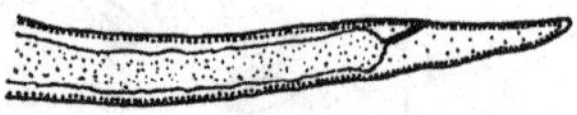
B. para-corneolus		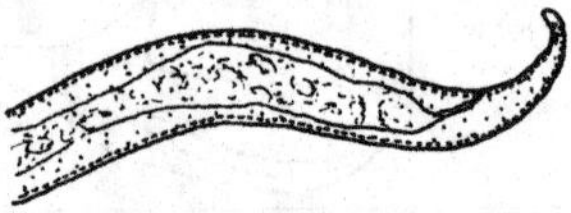
B. chitwoodi		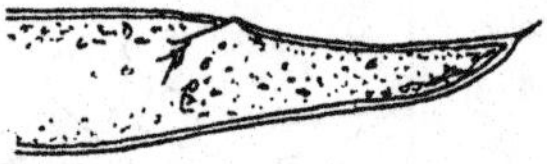
B. pinasteri	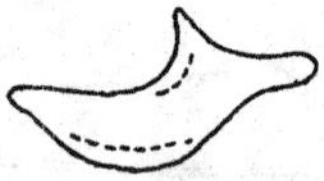	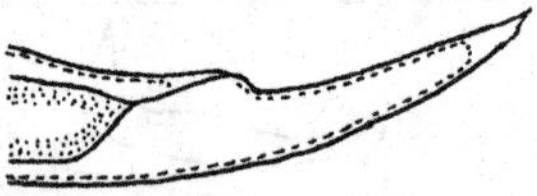
B. cryphali	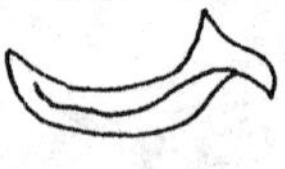	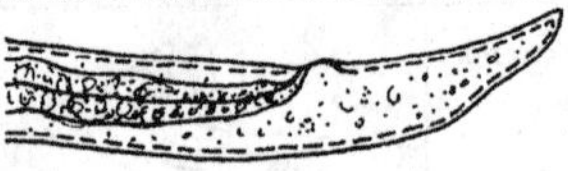
B. idius		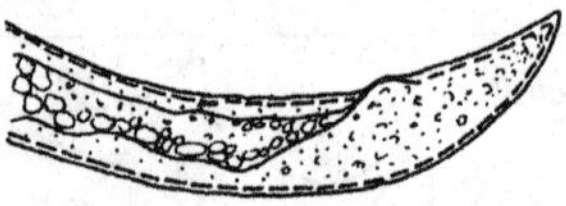
B. abietinus		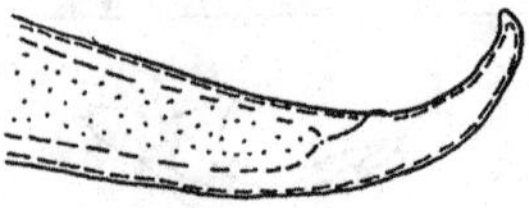
B. terato-spicularis		

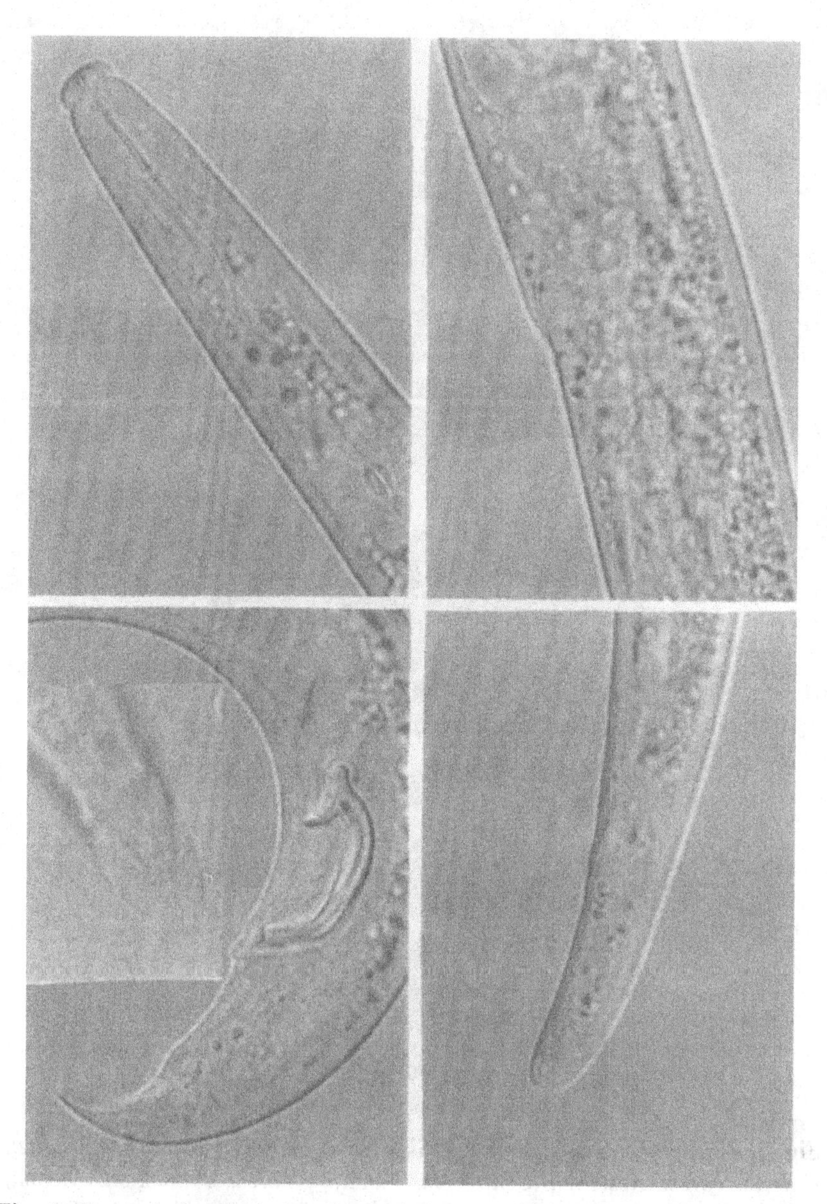

Fig. 1. Bursaphelenchus xylophilus *from Portugal (videoprinter microphotographs). Upper row: Anterior body and vulval region; Lower row: Male tail and female tail.*

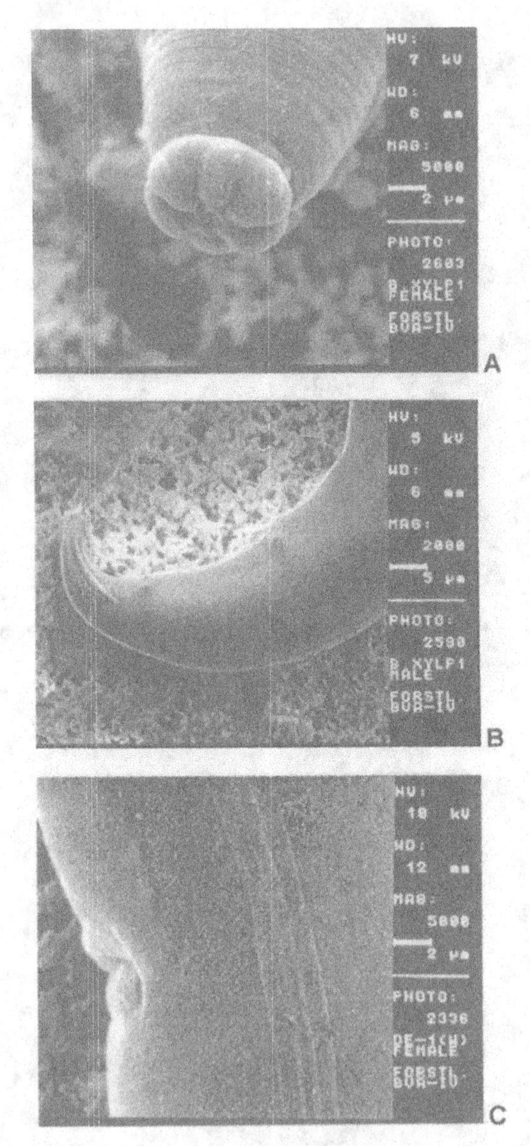

Fig. 2. Bursaphelenchus xylophilus *(SEM microphotographs M. Brandstetter, Federal Forest Research Centre, Institute for Forest Protection, Vienna, Austria). A: Female head; B: Male tail; C: Vulval region and lateral field.*

of them are not fully known. Species affiliation within the *B. sexdentati* group by morphology is, in some cases, also difficult due to the similarity of some species, whereas species of the *B. eggersi* group can be clearly

Table 3. *Measurements of* Bursaphelenchus xylophilus.

Character	Males			Females		
	Nickle *et al.*, 1981 (n = 5)	Mamiya & Kiyohara, 1972 (n = 30)	Mota *et al.*, 1999 (n = 12) (Portugal)	Nickle *et al.*, 1981 (n = 5)	Mamiya & Kiyohara, 1972 (n = 40)	Mota *et al.*, 1999 (n = 12) (Portugal)
Length (L), mm	0.56 (0.52-0.60)	0.73 (0.59-0.82)	1.03 (0.80-1.30)	0.52 (0.45-0.61)	0.81 (0.71-1.01)	1.05 (0.89-1.29)
a	40.8 (35-45)	42.3 (36-47)	49.4 (44-56)	42.6 (37-48)	40.0 (33-46)	50.0 (41-58)
b	9.4 (8.4-10.5)	9.4 (7.6-11.3)	13.3 (11.1-14.9)	9.6 (8.3-10.5)	10.3 (9.4-12.8)	13.8 (12.7-16.4)
c	24.4 (21-29)	26.4 (21-31)	28.0 (24-32)	27.2 (23-31)	26.0 (23-32)	26.6 (22-32)
Stylet, μm	13.3 (12.6-13.8)	14.9 (14-17)	12.6 (11-16)	12.8 (12.6-13.0)	15.9 (14-18)	12.3 (11-15)
Spicules, μm	21.2 (18.8-23.0)	27.0 (25-30)	24 (22-25)	–	–	–
Vulva position % of L	–	–	–	74.7 (73-78)	72.7 (67-78)	73.3 (70-76)

distinguished by spicule shape. *Bursaphelenchus leoni* can easily be distinguished from other species by combined consideration of the long, slim female tail and the characteristic spicule shape. The other two species of the group are known only from their original descriptions and were not studied in detail.

Within the *B. xylophilus* group, species can be morphologically distinguished by the shape of the female tail end. Spicules of *B. xylophilus, B. mucronatus* and *B. fraudulentus* are quite similar. The shape of their spicules is, however, very specific compared with all other *Bursaphelenchus* species. Therefore, affiliation of one of these three species to the *B. xylophilus* group can also be done by use of Table 2 and in case only males are found. *Bursaphelenchus xylophilus* has round-tailed females, *B. mucronatus* females have a clearly recognisable

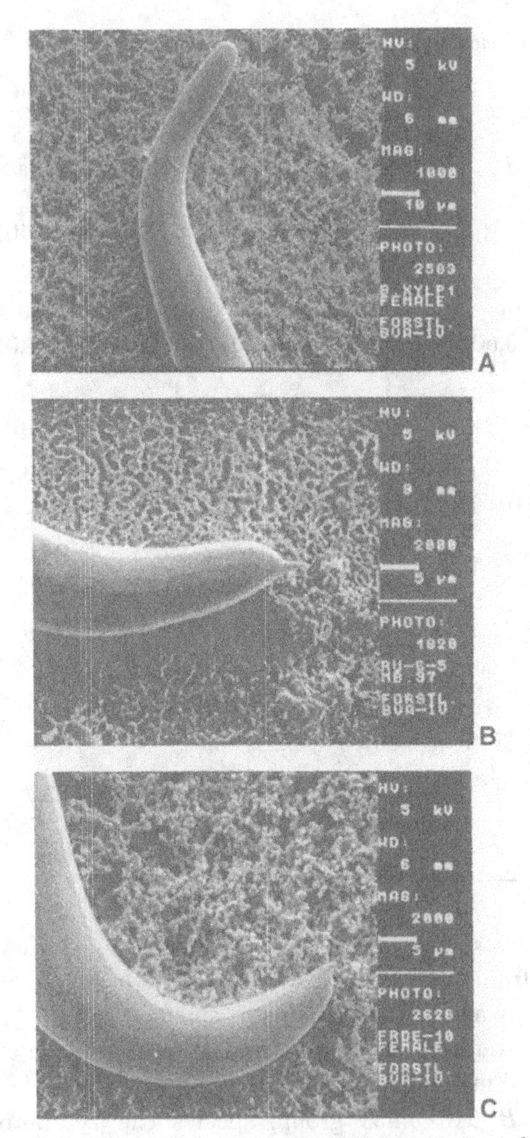

Fig. 3. *Shape of female tail ends (microphotographs M. Brandstetter, Federal Forest Research Centre, Institute for Forest Protection, Vienna, Austria). A:* Bursaphelenchus xylophilus*; B:* B. mucronatus*; C:* B. fraudulentus.

mucro, and *B. fraudulentus* has a female tail end shape in between the two other ones (Fig. 3). The mucronate form of *B. xylophilus*, which is morphologically similar to *B. mucronatus*, does not occur in

Europe. Specimens showing the typical features of the *B. xylophilus* group including the typical spicule shape and having a round female tail end can, therefore, be definitely identified as *B. xylophilus*. Occasionally, slightly mucronate forms may also be found in *B. xylophilus* populations with round-tailed females, for instance in Portugal, but these populations always have round-tailed females as well.

Bursaphelenchus mucronatus is also a species that shows high variability, particularly in the shape of the female tail end. A European and an East Asian genotype with slightly different mucro shapes and lengths (Braasch *et al.*, 1998) were differentiated by means of RAPD PCR and ITS-RFLP (Braasch *et al.*, 1995; Hoyer *et al.*, 1998). The East Asian genotype was found for the first time in Europe, in Germany. As far as it is known, however, populations of *B. mucronatus* never contain round-tailed females.

It is essential to use both males and females for identification. Whereas, in the case of *B. xylophilus*, the consideration of males alone allows at least affiliation to the *B. xylophilus* group, females may be confused with females of other *Bursaphelenchus* species. For instance, *B. sexdentati, B. naujaci, B. piniperdae* and *B. incurvus* are described as having round-tailed females. Vulval flaps of various size are present, for instance, in *B. sexdentati, B. naujaci, B. pinasteri* and *B. leoni*. Care must also be taken over identification by means of the spicules: the overall shape of the spicules has to be carefully considered, while a cucullus at the distal end of the spicules is present not only in the case of *B. xylophilus, B. mucronatus* and *B. fraudulentus*, but also with other *Bursaphelenchus* species such as *B. hellenicus, B. hofmanni, B. para-corneolus, B. abietinus* and *B. pinophilus*.

JUVENILES

Morphological assignment of juveniles to species is difficult in most cases. The second stage juveniles (J2) are round-tailed and similar in all species. Special characters of the tail can sometimes be seen in the third stage, whereas the fourth juvenile stage (J4) already shows a similar tail shape to the females. *Bursaphelenchus xylophilus* J4 have a round tail end, *B. mucronatus* J4 a distinct mucronate tail end and *B. fraudulentus* J4 a tail end in between the two others. Therefore, the presence of typical males of the group and of J4 may already give a valuable clue about species identification.

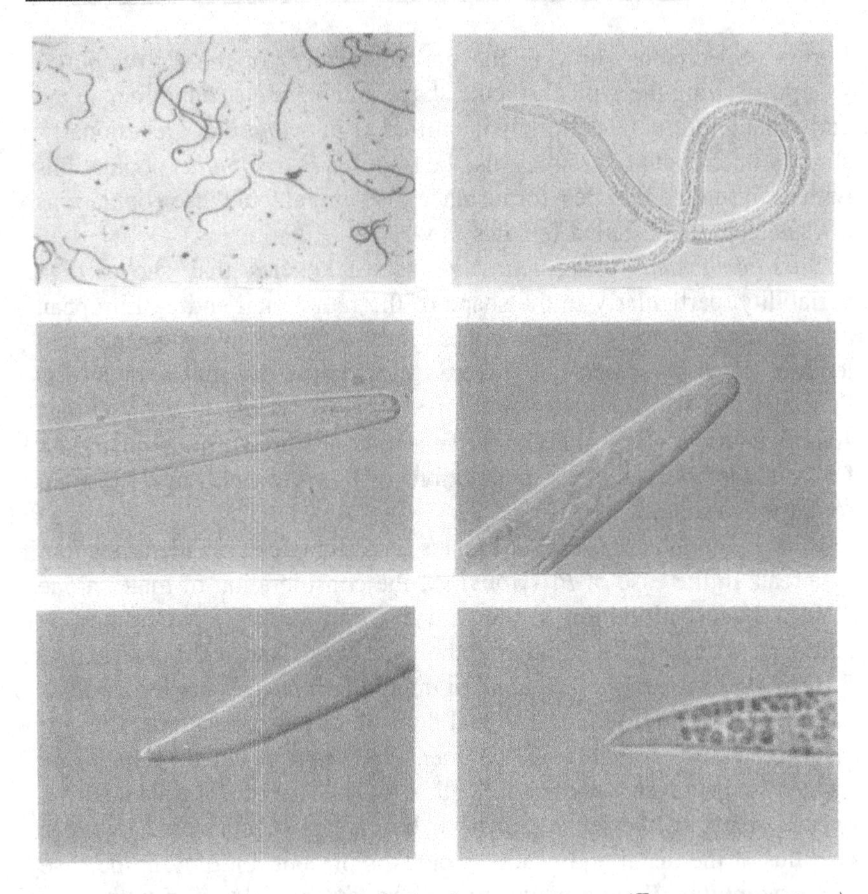

Fig. 4. *Dauer juveniles of* Bursaphelenchus mucronatus *(European genotype). Upper row: Whole body; Middle row: Head end; Lower row: Tail end.*

Dauer juveniles found on the vector beetles have a dome-shaped head, which is not offset from the body. Pharynx and median bulb are poorly defined, and the body is full of contents. The tail of both *B. xylophilus* and *B. mucronatus* is conoid and sharply pointed and with distinct mucro in the case of *B. mucronatus* (Figs 4, 5).

Discussion

Bursaphelenchus xylophilus is characterised and can clearly be morphologically identified in a European survey by the typical shape of spicules with cucullus and the terminal bursa of males, and by the

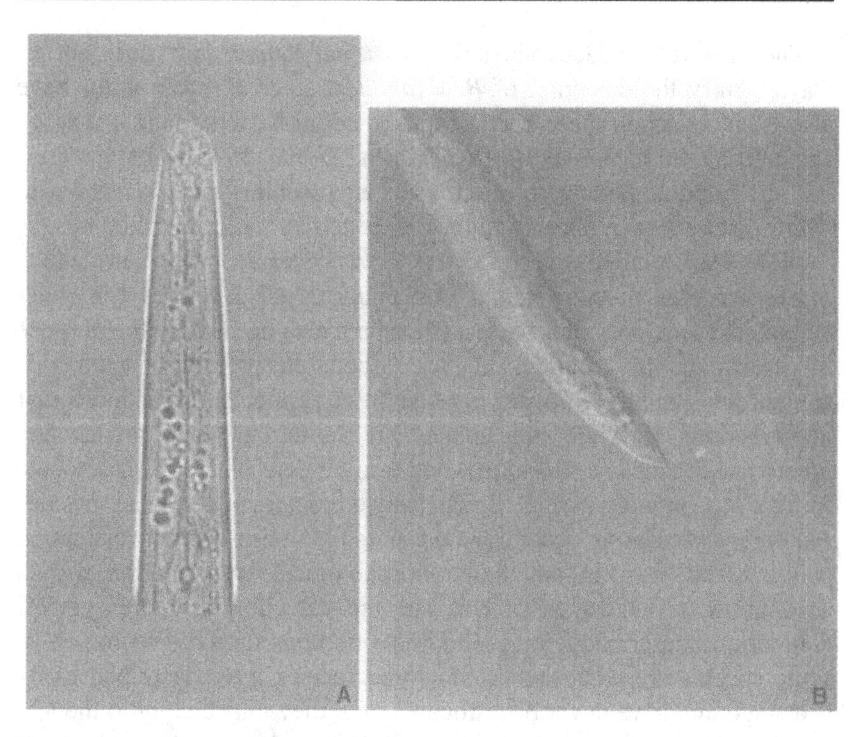

Fig. 5. *Dauer juvenile of* Bursaphelenchus xylophilus *(provided by C. Penas, Portugal). A: Head; B: Tail.*

presence of a distinct vulval flap and a round tail terminus of females. *Bursaphelenchus xylophilus* strains with mucronate tail terminus, *i.e.*, with distinct mucro; have not been found outside North America. Confusion is possible due to the similar spicule shape in *B. mucronatus* and *B. fraudulentus*, the presence of a cucullus and a vulval flap in other *Bursaphelenchus* species, and the occurrence of females with a vulval flap and with rounded tail terminus in other tree-inhabiting genera. Sometimes, the female tail terminus of *B. xylophilus* shows a tiny mucro. It is, in any case, shorter than that of *B. mucronatus*, but may be a reason to mistake it for *B. fraudulentus*, which also has a short mucro. It has also been observed that the percentage of *B. xylophilus* specimens with a small mucro in a population increases after inoculation of laboratory-cultured nematodes into host plants. Round-tailed females are, however, always present in those *B. xylophilus* populations. For the reliable morphological identification of *B. xylophilus*, males and females should always be considered. The presence of J4 may be helpful.

The discovery of dauer juveniles on *Monochamus* spp. does not always signify the detection of *B. xylophilus*, since *B. mucronatus* have also been found on these beetles and, in addition, other nematodes are vectored by *Monochamus* (*Bradynema* sp., Poinar, 1975; *Diplogasteroides* sp., unpubl.). For confirmation, dauer juveniles must be reared to adults on *Botrytis* or another fungus, or tested by molecular methods.

Molecular methods need to be used in cases where no adults of *Bursaphelenchus* are found and culturing of juveniles (or dauer juveniles) is not possible. Such methods are also needed for the purpose of confirming new findings in areas where the pine wood nematode has not been known to occur previously. A further need for molecular investigations is when only males are found, or when males and mucronate females are found in wood from regions where the mucronate form of *B. xylophilus* occurs. Since many other nematode species besides *Bursaphelenchus* spp. occur in wood, an initial microscopic investigation is warranted, and, in any case, morphological identification should accompany molecular tests. Besides, the detection of more species, including undescribed species, and new findings on the distribution of some species are to be expected. Some species were described many years ago and have never been found again; they may belong to the less common species. Molecular patterns are not available in these cases.

It should also be remarked that a few species very similar to members of the *B. xylophilus* group (*B. abruptus* Giblin-Davis, Mundo-Ocambo, Baldwin, Norden & Batra, 1993; *B. conicaudatus* Kanzaki, Tsuda & Futai, 2000) have been described; they are not, however, known to occur in conifers. *Bursaphelenchus abruptus* is associated with a digger bee in North America, whereas *B. conicaudatus* was found in fig trees in Japan. *Bursaphelenchus kolymensis* Korenchenko, 1980 found in Russia is considered to be synonymous with *B. mucronatus*. Further detailed taxonomic work is necessary in order to improve the reliability of identification of less frequent *Bursaphelenchus* species and of some species very similar and possibly synonymous to each other, such as in the *B. sexdentati* group. In the case of *B. xylophilus*, however, the taxonomic position and the morphology are quite clear.

References

BRAASCH, H. (2001). *Bursaphelenchus* species in conifers in Europe: distribution and morphological relationship. *EPPO Bulletin* 31, 127-142.

BRAASCH, H., BURGERMEISTER, W., HARMEY, M.A., MICHALOPOULOS-SKARMOUTSOS, H., TOMICZEK, C. & CAROPPO, S. (2000). Pest risk analysis of pinewood nematode related *Bursaphelenchus* species in view of South European pine wilting and wood imports from Asia. *Final Report of EU research project*, Fair CT 95-0083, 251 pp.

BRAASCH, H., BURGERMEISTER, W. & HOYER, U. (1998). Diversity within the species *Bursaphelenchus mucronatus* worldwide. *Nematologica* 44, 465-466.

BRAASCH, H., BURGERMEISTER, W. & PASTRIK, K.-H. (1995). Differentiation of three *Bursaphelenchus* species by means of RAPD-PCR. *Nachrichtenblatt des Deutschen Pflanzenschutzdienstes* 47, 310-314.

HOYER, U., BURGERMEISTER, W. & BRAASCH, H. (1998). Identification of *Bursaphelenchus* species (Nematoda, Aphelenchoididae) on the basis of amplified ribosomal DNA (ITS-RFLP). *Nachrichtenblatt des Deutschen Pflanzenschutzdienstes* 50, 273-277.

MOTA, M.M., BRAASCH, H., BRAVO, M.A., PENAS, A.C., BURGERMEISTER, W., METGE, K. & SOUSA, E. (1999). First record of *Bursaphelenchus xylophilus* in Portugal and in Europe. *Nematology* 1, 727-734.

NICKLE, W.R., GOLDEN, A.M., MAMIYA, Y. & WERGIN, W.P. (1981). On the taxonomy and morphology of the pine wood nematode, *Bursaphelenchus xylophilus* (Steiner & Buhrer, 1934) Nickle, 1970. *Journal of Nematology* 13, 385-392.

POINAR, G.O. (1975). *Entomogenous nematodes*. Leiden, The Netherlands, Brill, 317 pp.

SCHÖNFELD, U., APEL, K.-H. & BRAASCH, H. (2001). Nematoden der Gattung *Bursaphelenchus* (Nematoda, Parasitaphelenchidae) in den Kiefernwäldern des Landes Brandenburg – Ergebnisse eines Monitoring. *Nachrichtenblatt des Deutschen Pflanzenschutzdienstes* 53, 180-184.

YIK, C.-P. & BIRCHFIELD, W. (1981). Observations on the morphology of the pine wood nematode, *Bursaphelenchus xylophilus*. *Journal of Nematology* 13, 376-384.

Nematology Monographs & Perspectives, 2003, Vol. 1, 145-153

Molecular evidence for *Bursaphelenchus xylophilus* x *B. mucronatus* hybridisation under experimental conditions

Wolfgang BURGERMEISTER [1] and Helen BRAASCH [2]

[1] *Federal Biological Research Centre for Agriculture and Forestry, Institute for Plant Virology, Microbiology and Biosafety, Messeweg 11, 38104 Braunschweig, Germany*

[2] *Federal Biological Research Centre for Agriculture and Forestry, Department for National and International Plant Health, Kleinmachnow Branch, Stahnsdorfer Damm 81, 14532 Kleinmachnow, Germany*

Summary – *Bursaphelenchus* species can be differentiated by means of ITS-RFLP analysis. In this technique, an rDNA fragment containing the ITS-1 and ITS-2 regions is amplified and analysed for restriction fragment length polymorphisms. Using five restriction enzymes, we have established specific fragment patterns for 17 *Bursaphelenchus* species so far. Sufficient DNA for a complete analysis can be obtained from a single nematode, and individual animals of the same species have been shown to produce identical species-specific ITS-RFLP patterns. Experimental hybridisation of *B. xylophilus* and *B. mucronatus* was attempted by co-culturing of both species on *Botrytis cinerea* using five malt agar plates in Petri dishes (initial number 50 specimens of each species) for each trial and by co-inoculation of 3000 specimens each per plant onto five 4-year-old *Pinus sylvestris* per trial. In both methods, a *B. xylophilus* strain from Portugal was combined with a European genotype of *B. mucronatus* on the one hand, and with an East Asian genotype of *B. mucronatus* on the other hand. Both strains of *B. mucronatus* originated in Germany. The Petri dishes and plants were kept at 25°C in climate chambers. All inoculated plants died within 7 weeks. Nematodes reisolated from wilted plants or from culture plates were inspected under a light-microscope, and females were classified according to the shape of their tail terminus as typical for *B. xylophilus*, *B. mucronatus* or intermediates. The same specimens were then subjected to ITS-RFLP analysis. Among the nematodes reisolated from inoculated trees, 19 out of 29 specimens examined to date exhibited RFLP patterns corresponding to *B. xylophilus*. Six specimens were *B. mucronatus*, and the remaining four specimens exhibited intermediate RFLP patterns containing typical fragments of *B. xylophilus* and *B. mucronatus*. Among the nematodes reisolated from culture plates, 22 specimens have been examined by ITS-RFLP so far. Of these, 14 specimens were classified as *B. xylophilus* and four specimens as *B. mucronatus*. The remaining four specimens exhibited intermediate patterns containing typical fragments of both species to varying degrees. The results indicate that hybridisation of *B. xylophilus* × *B. mucronatus* (European and East Asian genotypes) and backcrossing had occurred under the experimental conditions described. More specimens have to be examined to see if intermediate morphological features are correlated with intermediate RFLP patterns of possible hybrid specimens.

Bursaphelenchus xylophilus and *B. mucronatus* are closely related species, as has been concluded from the striking similarities in their host plants and vector preference, life cycle and morphological features. In areas where both species occur, hybrid formation may be possible with unknown consequences for phytopathology. Reports on successful experimental hybridisation have been published by several nematologists during the last 15 years. Reciprocal crosses were made by de Guiran and Bruguier (1989) between four strains of *B. xylophilus* and two strains of *B. mucronatus*. The French *B. mucronatus* strain gave fertile hybrids with both *B. mucronatus* from Japan and *B. xylophilus* from Japan. In trials of Riga *et al.* (1992), interspecific hybridisations among isolates of *B. mucronatus* from France and Japan and *B. xylophilus* from North America and Japan were non-reciprocal in some cases, *i.e.*, the F1 died out in most of the crosses in which the female was *B. mucronatus*. Bolla and Boschert (1993) found that interbreeding of *B. xylophilus* and *B. mucronatus* was rare, although under laboratory conditions interbreeding can be forced. In crossing experiments of Schauer-Blume (1992), only the French isolate of *B. mucronatus* gave rise to a fertile F1 progeny with *B. xylophilus* from Japan, whereas the F1 produced by *B. mucronatus* from Norway and *B. xylophilus* from Japan was sterile. A German isolate of *B. mucronatus* produced, however, a fertile F1 with *B. xylophilus* from Japan, while crossings among a *B. mucronatus* strain from Russia and the same *B. xylophilus* strain were less successful (Braasch, 1994). It may be concluded from laboratory experiments that hybridisation between the two species is possible, and that the results of hybridisation experiments depend much on the strains used and are not always reciprocal. Besides, the results may be influenced by experimental conditions. It is, however, still unclear whether gene flow in the wild does exist.

In view of the recent discovery of *B. xylophilus* in Portugal (Mota *et al.*, 1999) and the widespread occurrence of *B. mucronatus* in Europe, we have carried out a new series of hybridisation experiments, using a Portuguese *B. xylophilus* strain for the first time, and following two different experimental approaches. In the first approach, *B. xylophilus* and *B. mucronatus* were co-cultured on *Botrytis cinerea* malt agar plates in Petri dishes. In the second approach, the two species were co-inoculated onto 4-year-old seedlings of *Pinus sylvestris*. After appropriate times, nematodes were reisolated and investigated by light-microscopy and ITS-RFLP analysis.

The molecular technique of ITS-RFLP analysis may offer a convenient method for identification of hybrids. In this technique, total DNA is extracted from individual nematodes, and a fragment of rDNA including the ITS-1 and ITS-2 sequences is amplified by PCR. The amplified rDNA is then digested with restriction enzymes, and species-specific subfragments are obtained which are separated by agarose gel electrophoresis and visualised as ITS-RFLP patterns. In previous work, ITS-RFLP patterns have been established for identification of 17 *Bursaphelenchus* species so far (Braasch *et al.*, 1999, 2001), and individual animals of the same species have been shown to produce identical species-specific ITS-RFLP patterns.

Materials and methods

In the first trial, 50 specimens of each *B. xylophilus* and *B. mucronatus* were co-cultured on a *Botrytis cinerea* malt agar plate in a Petri dish and kept at 25°C for 2-3 weeks. In the second trial, 3000 specimens of each species were co-inoculated onto a 4-year-old seedling of *Pinus sylvestris*, following a published procedure (Braasch, 1997).

In both methods, a *B. xylophilus* strain from Portugal (PT-1w) was combined with a European genotype of *B. mucronatus* (DE-3w) on the one hand, and with an East Asian genotype of *B. mucronatus* (DE-5w) on the other hand. Both strains of *B. mucronatus* originated in Germany.

All experiments were carried out with five replications. Therefore, a total number of ten Petri dishes and ten pine seedlings were employed in the hybridisation experiments.

The inoculated plants were kept at 25°C in climate chambers. All inoculated plants died within 7 weeks. Nematodes from wilted plants or from culture plates were reisolated and inspected under light-microscope, and females were classified according to the shape of their tail terminus as typical for *B. xylophilus* (round-tailed), *B. mucronatus* (mucronate) or intermediate forms. For molecular investigation, individual specimens were homogenised, DNA was extracted and ITS-RFLP analysis was carried out as described previously (Hoyer *et al.*, 1998).

Results and discussion

The results of morphological inspection are presented in Table 1. Round-tailed specimens as well as specimens with varying length of mucro were found, thus indicating that hybridisation between the two species had probably occurred. The situation is, however, complicated by the fact that *B. xylophilus*, particularly if isolated from trees, may also have a short mucro in some cases. The portion of females with a finger-like short mucro (different from the typical long mucro of *B. mucronatus*) in the progeny was considerably high. Since most of the single females examined by ITS-RFLP (see below) were shown to be *B. xylophilus*, a large portion of the specimens with a short mucro obviously belonged to *B. xylophilus*.

In ITS-RFLP analysis, restriction fragment patterns of individual nematodes reisolated from hybridisation experiments were compared with reference patterns established for *B. xylophilus* and *B. mucronatus*, respectively (Mota *et al.*, 1999). Some specimens exhibited a mixed type of ITS-RFLP pattern, as shown in Fig. 1. Most bands of the specimen shown in the centre can either be assigned to the reference pattern of *B. xylophilus* or *B. mucronatus* shown at both sides. This suggests that the specimen is a product of hybridisation between the two species. Another example of a mixed-type ITS-RFLP pattern is shown in Fig. 2. Again, many bands can be assigned to either parent species. Interestingly, the presumably hybrid patterns in Figs 1 and 2 cannot be explained by a simple superimposition of the two reference patterns, because additional bands are also visible which cannot be ascribed to either of the parent species. This is also illustrated in Fig. 3 where the ITS-RFLP patterns of three presumably hybrid specimens are shown side by side. Obviously, these patterns are all complex but not identical.

Table 1. *Proportion of females with different tail end shapes (r: round; b: with 1 μm blunt projection; s: with short finger-like mucro; l: with long tapering mucro) in the progeny of* Bursaphelenchus xylophilus *(Bx) and two genotypes (E: European; EA: East Asian) of* B. mucronatus *(Bm) co-inoculated onto* Botrytis cinerea *(Bc) or 4-year-old* Pinus sylvestris *(Ps).*

Bursaphelenchus isolates co-inoculated	Nutrition	Number of specimens inoculated (Bx/Bm)	Number of females examined	Number (%) females in four categories			
				r	b	s	l
Bx & Bm (E)	Bc	50/50 per plate	50	17 (34)	4 (8)	23 (46)	6 (12)
	Ps	3000/3000 per tree	37 (all available females)	7 (19)	1 (3)	15 (40)	14 (38)
Bx & Bm (EA)	Bc	50/50 per plate	50	7 (14)	0 (0)	15 (30)	28 (56)
	Ps	3000/3000 per tree	250 (50 from each of 5 trees)	3 (1)	14 (6)	173 (69)	60 (24)

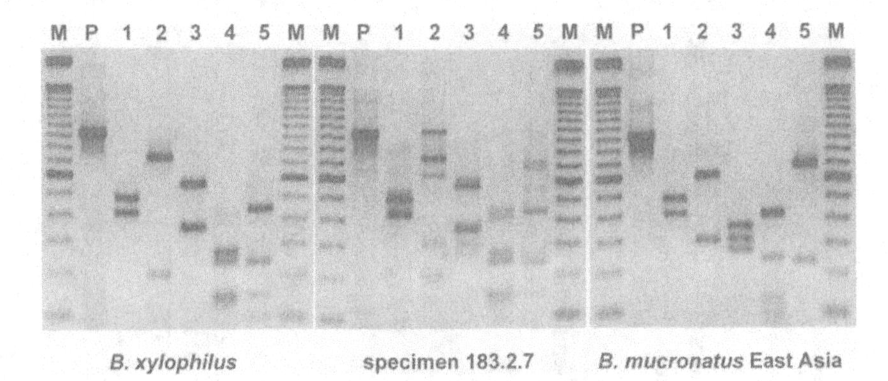

Fig. 1. *ITS-RFLP patterns of* Bursaphelenchus xylophilus, B. mucronatus *East Asia and a possible hybrid (specimen 183.2.7) obtained after co-culture of both species on* Botrytis cinerea *agar. rDNA amplification products (P) and their restriction fragments obtained with* Rsa*I (1),* Hae*III (2),* Msp*I (3),* Hinf*I (4) and* Alu*I (5). M: 100 bp DNA ladder.*

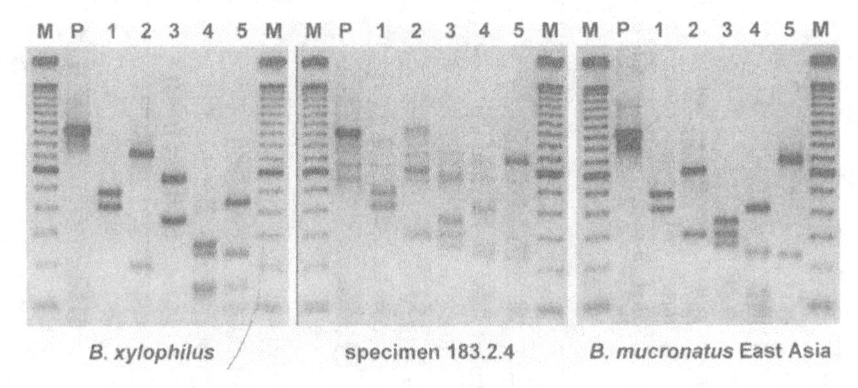

Fig. 2. *ITS-RFLP patterns of* Bursaphelenchus xylophilus, B. mucronatus *East Asia and a possible hybrid (specimen 183.2.4) obtained after co-culture of both species on* Botrytis cinerea *agar. rDNA amplification products (P) and their restriction fragments obtained with* Rsa*I (1),* Hae*III (2),* Msp*I (3),* Hinf*I (4) and* Alu*I (5). M: 100 bp DNA ladder.*

There are differences in the number and intensity of bands. It should be kept in mind that the specimens examined are not the product of a controlled 1:1 hybridisation but the result of prolonged co-cultivation of both species. Therefore, backcrossing of hybrids with either parent species has to be considered. As a result, hybrid offspring with different genome share of *B. xylophilus* and *B. mucronatus* can be expected and

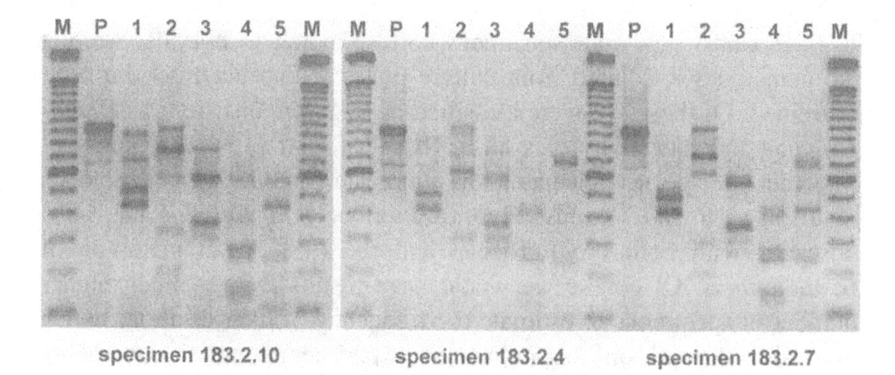

Fig. 3. *ITS-RFLP patterns of possible hybrids obtained during co-culture of* Bursaphelenchus xylophilus *and* B. mucronatus *East Asia on* Botrytis cinerea *agar. rDNA amplification products (P) and their restriction fragments obtained with* Rsa*I (1),* Hae*III (2),* Msp*I (3),* Hinf*I (4) and* Alu*I (5). M: 100 bp DNA ladder.*

Table 2. *Results of ITS-RFLP analysis of nematodes reisolated from hybridisation experiments.*

Hybridisation experiment	Method	Number of specimens examined	Number of *B. xylophilus* found	Number of hybrids found	Number of *B. mucronatus* found
B. xylophilus and	*Botrytis* agar plate	9	7	1	1
B. mucronatus (European genotype)	*Pinus sylvestris* seedling	4	4	0	0
B. xylophilus and	*Botrytis* agar plate	13	7	3	3
B. mucronatus (East Asian genotype)	*Pinus sylvestris* seedling	25	15	4	6

may be responsible for the varying ITS-RFLP patterns that have been obtained.

The ITS-RFLP results obtained so far are summarised in Table 2. Among the nematodes reisolated from inoculated trees, 29 specimens have been examined. Of these, 19 specimens exhibited ITS-RFLP patterns corresponding to *B. xylophilus*, six specimens corresponding to

B. mucronatus, and four specimens corresponding to hybrids. Among the nematodes reisolated from culture plates, 22 specimens have been examined. Of these, 14 were classified as *B. xylophilus*, four as *B. mucronatus* and four as hybrids on the basis of their ITS-RFLP patterns. The ratio of species reisolated may indicate that *B. xylophilus* tends to displace *B. mucronatus*. Hybridisation of *B. xylophilus* has been observed with both the European and the East Asian genotype of *B.mucronatus*. Of course we would need results from a much higher number of specimens to estimate the frequency of hybridisation, which may also depend on the individual experimental conditions. More specimens also need to be examined in order to see if intermediate morphological features are correlated with intermediate ITS-RFLP patterns of hybrid specimens. At this stage, we may only conclude that ITS-RFLP findings have revealed that hybridisation has occurred during co-culture on plates as well as after co-inoculation onto plants. It will be interesting to see whether hybrids can also be found outside the laboratory in areas where *B. xylophilus* and *B. mucronatus* coexist.

Acknowledgement

We would like to express our gratitude to Brigitte Toenhardt and Elvira Buchbach who have provided excellent technical assistance in the nematological and molecular biology experiments.

References

BOLLA, R.I. & BOSCHERT, M. (1993). Pinewood nematode species complex: interbreeding potential and chromosome number. *Journal of Nematology* 25, 227-238.

BRAASCH, H. (1994). [Intra- and interspecific crossing experiments with a German and a Siberian isolate of *Bursaphelenchus mucronatus* (Nematoda, Aphelenchoididae) and related *Bursaphelenchus* species.] *Nachrichtenblatt des Deutschen Pflanzenschutzdienstes* 46, 276-281.

BRAASCH, H. (1997). [Investigations on host plants and pathogenicity of the pine wood nematode (*Bursaphelenchus xylophilus*) from North America under Central European climate conditions.] *Nachrichtenblatt des Deutschen Pflanzenschutzdienstes* 49, 209-214.

BRAASCH, H., METGE, K. & BURGERMEISTER, W. (1999). [*Bursaphelenchus* species (Nematoda, Parasitaphelenchidae) found in coniferous trees in Germany and their ITS-RFLP patterns.] *Nachrichtenblatt des Deutschen Pflanzenschutzdienstes* 51, 312-320.

BRAASCH, H., TOMICZEK, C., METGE, K., HOYER, U., BURGERMEISTER, W., WULFERT, I. & SCHÖNFELD, U. (2001). Records of *Bursaphelenchus* spp. (Nematoda, Parasitaphelenchidae) in coniferous timber imported from the Asian part of Russia. *Forest Pathology* 31, 129-140.

DE GUIRAN, G. & BRUGUIER, N. (1989). Hybridization and phylogeny of the pine wood nematode (*Bursaphelenchus* spp.). *Nematologica* 35, 321-330.

HOYER, U., BURGERMEISTER, W. & BRAASCH, H. (1998). Identification of *Bursaphelenchus* species (Nematoda, Aphelenchoididae) on the basis of amplified ribosomal DNA (ITS-RFLP). *Nachrichtenblatt des Deutschen Pflanzenschutzdienstes* 50, 273-277.

MOTA, M.M., BRAASCH, H., BRAVO, M.A., PENAS, A.C., BURGERMEISTER, W., METGE, K. & SOUSA, E. (1999). First report of *Bursaphelenchus xylophilus* in Portugal and in Europe. *Nematology* 1, 727-734.

RIGA, E., BECKENBACH, K. & WEBSTER, J.M. (1992). Taxonomic relationship of *Bursaphelenchus xylophilus* and *B. mucronatus* based on interspecific and intraspecific cross-hybridization and DNA analysis. *Fundamental and Applied Nematology* 15, 391-395.

SCHAUER-BLUME, M. (1992). Crossing experiments with *Bursaphelenchus* isolates from Europe and Japan. *Zeitschrift für Pflanzenkrankheiten und Pflanzenschutz* 99, 304-310.

Nematology Monographs & Perspectives, 2003, Vol. 1, 155-163

Satellite DNA used as a species specific probe for identification of the pine wood nematode *Bursaphelenchus xylophilus*

Pierre ABAD

INRA, Unité de Santé Végétale et Environnement, BP2078, 06606 Antibes cedex, France

Summary – We developed simple and efficient methods to identify *B. xylophilus* isolates based upon satellite DNA sequences, through PCR and squashed nematode experiments. Using the *Msp*I satellite DNA, these methods appeared to be specific to this nematode species, and were also highly sensitive since detection of a single individual was achieved. Therefore, in the pine wood nematode species complex, identification of the plant-pathogenic species *B. xylophilus* may be carried out both rapidly and reliably. This step may be crucial for the success of direct detection of *B. xylophilus* from wood samples, and will prove to be of great use in ecological and population studies.

The pine wood nematode, *Bursaphelenchus xylophilus* is the causal agent of pine wilt disease. The identification of *B. xylophilus* is difficult because of its similarity to *B. mucronatus* and *B. fraudulentus* which are not pathogenic on pine trees under field conditions (Mamiya & Enda, 1979; Kiyohara & Bolla, 1990). *Bursaphelenchus mucronatus* differs from *B. xylophilus* only by the presence of a mucron on the female tail. However, a North American *B. xylophilus* isolate, US10, is morphologically similar to *B. mucronatus*. These three species are thought to be derived from a common origin and, therefore, it has been proposed that *B. xylophilus*, *B. mucronatus* and *B. fraudulentus* constitute a supraspecies referred as 'the pinewood nematode species complex' (PWNSC) (de Guiran & Bruguier, 1989). In this context it is important to have some means of identifying *B. xylophilus*, and there is an urgent need for simple and reliable tools for identification of these *Bursaphelenchus* species. Attempts have been made to distinguish between these nematodes by using morphological characteristics and enzyme electrophoresis (de Guiran *et al.*, 1985). However, they failed to solve the problem because of the strong conservation of proteins between closely related taxa, and because of the modification of their expression by environmental factors. Therefore, genotyping-based detection systems using different molecular markers have been carried out (Webster *et al.*, 1990; Tarès *et al.*, 1992; Harmey & Harmey, 1993). This article reports the main results of research carried out at INRA Antibes, France, in order to develop new molecular tools based on repetitive sequences, know as satellite DNA, for *B. xylophilus* identification.

THE *MSP*I SATELLITE DNA FROM *B. XYLOPHILUS*

Satellite DNA is present in the genome of almost all eukaryotic organisms and is composed of non-coding highly repetitive sequences organised as long arrays of tandemly repeated elements, present as up to 100 000 copies per haploid genome. Depending on the organisms tested, these sequences represent less than 1% to more than 70% of the genome. Each family differs by the length of the monomer (70-2000 bp) or its sequence. In a number of cases (including nematodes), satellite DNA have been found to be species specific, even between sibling species. Despite extensive attempts to elucidate their biological role, no clear conclusion has been obtained with regard to any possible function. Because these sequences are often located at centromeric and telomeric regions of the chromosomes, it has been postulated that they can be involved in chromosome pairing.Therefore, satellite DNA has been considered of potential interest for nematode diagnostics because of the particular features mentioned above (high copy number and species specificity).

Firstly, we isolated a highly repetitive DNA sequence from *B. xylophilus*. It is represented as a tandemly repeated *Msp*I site-containing sequence with a monomeric unit of 160 bp and constitutes up to 30% of the nematode genome. The consensus sequence is 62% A+T rich, with the presence of direct, indirect and invert repeat clusters. Analysis of monomer sequences shows an average divergence of 3.9% from the calculated consensus. The *Msp*I satellite sequence is species specific since it hybridises only with genomic DNA from the *B. xylophilus* isolates (Tarès *et al.*, 1993).

The variability of this satellite family has been used to demonstrate its usefulness in fingerprinting since it generates random fragment length polymorphism (RFLP) between the *B. xylophilus* isolates. In Southern blotting experiments, we showed that the American and Japanese isolates hybridise more strongly to the *Msp*I probe than do the Canadian isolates. Therefore, Japanese isolates of *B. xylophilus* seem to be closer to the USA ones than to the Canadian isolates. This result is in agreement with those previously obtained with other probes, and supports the hypothesis that a *B. xylophilus* isolate reached Japan from North America probably from the United States (Tarès *et al.*, 1992). Concerning the status of the French populations, these results clearly show that they do not belong to the pathogenic species *B. xylophilus.*

Satellite DNA as a useful tool for diagnostic purposes

Because of its abundance and its species specificity, the *Msp*I family has been developed as a sensitive and reliable probe for the direct identification of *B. xylophilus* from wood samples (Tarès *et al.*, 1994).

Squashed nematode experiment

First, we developed a 'squashed nematode experiment' which allows the detection of a single pathogenic *B. xylophilus* nematode and we defined experimental conditions for a non radioactive identification method.

To determine whether the *Msp*I satellite could be used as a rapid diagnostic probe, we devised a simple direct hybridisation method on nematode prints (the nematodes are crushed by gentle pressure with a pipette tip onto a nylon membrane). We showed that the *Msp*I satellite DNA probe was effective directly on squashed nematodes spotted onto a filter using a very simple procedure with no need to extract DNA. The tested populations of *B. xylophilus* gave a strong signal while the *B. mucronatus* and *B. fraudulentus* populations never hybridised with the *Msp*I satellite DNA labelled with radioactive phosphorus. The use of this *Msp*I satellite DNA as a species specific probe leaves no doubt as to the identity of the nematode.

However, the only limitation of using this method on a routine basis is that it requires radioactively-labelled compounds. The development of a non radioactive identification method is a crucial step for diagnostic purposes routinely used by the Plant Protection Services in Europe. Therefore, next we developed a non radioactive identification method using a digoxigenin (DIG) labelled probe for 'squashed nematode

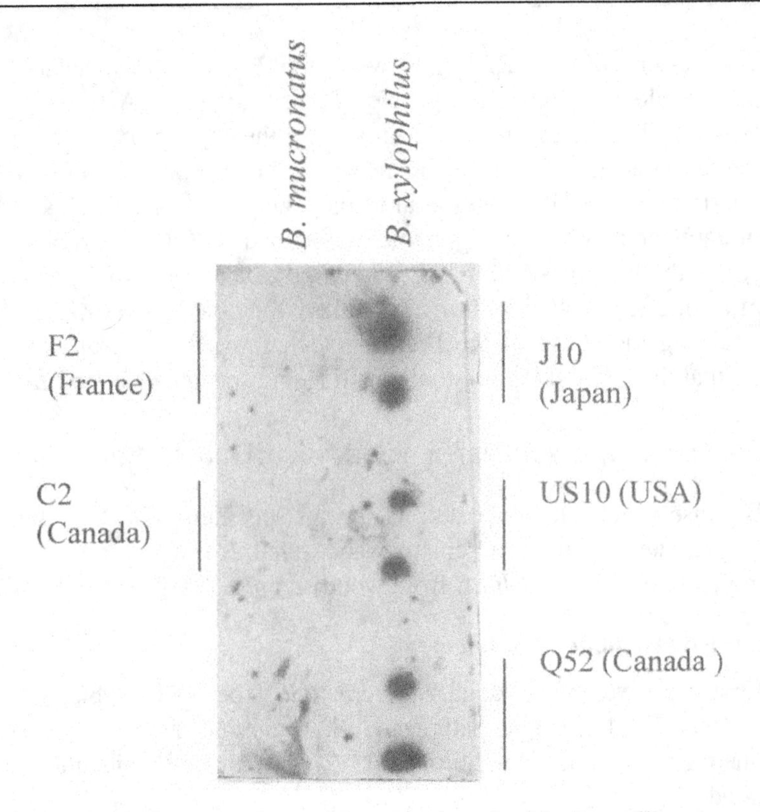

Fig. 1. *Squashed nematode experiment using the* Msp*I satellite as probe. The digoxigenin (DIG) labelled probe for 'squashed nematode experiment' allowing the detection of a single pathogenic nematode* Bursaphelenchus xylophilus. *The tested populations of* B. xylophilus *gave a strong signal while the* B. mucronatus *and* B. fraudulentus *populations never hybridised.*

experiment' allowing the detection of a single pathogenic nematode *B. xylophilus* (Fig. 1).

PCR detection procedure

Second, we developed a PCR method which allows specific amplification of DNA sequence from *B. xylophilus* populations. We defined the optimal PCR conditions to get amplification from a single worm alone or in a mixture with other *Bursaphelenchus* or Rhabditidae species.

Since its development, PCR technology has become an extremely powerful tool for plant disease diagnosis. Because satellite DNA sequences have often been found to be species specific, choosing a

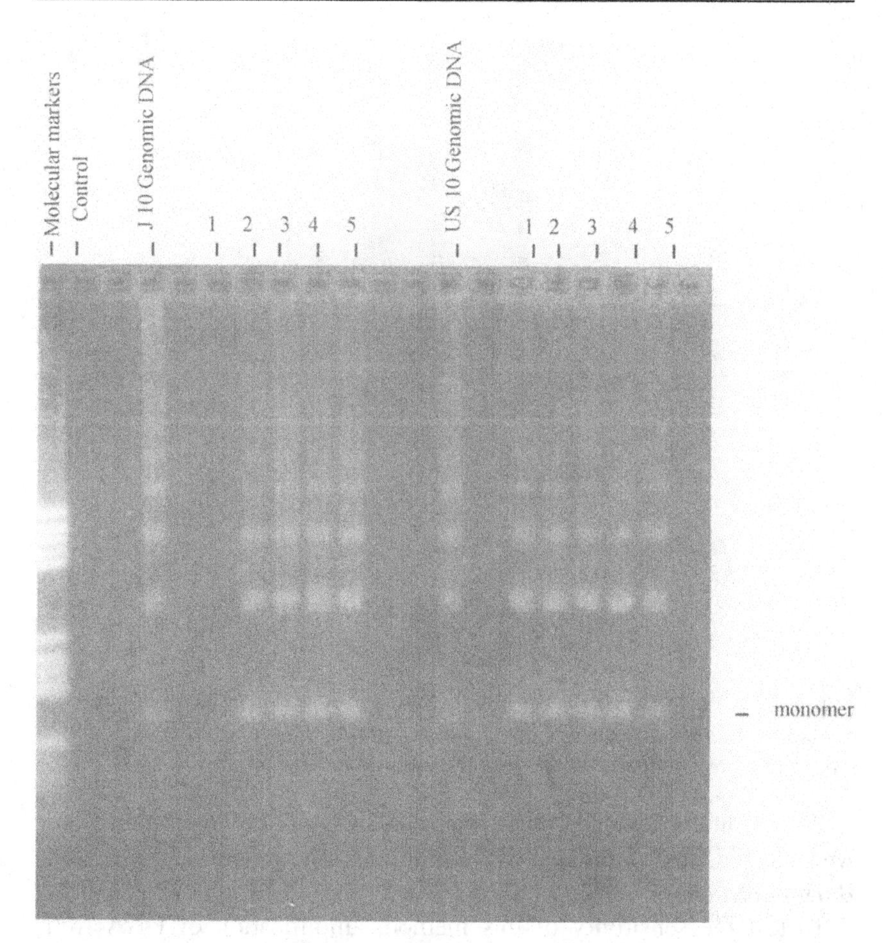

Fig. 2. *PCR experiments from* Bursaphelenchus xylophilus *single adults. Individual amplification was obtained using primers deduced from the sequence of the monomeric unit of the* MspI *satellite DNA. Amplification patterns from individuals were always very similar to the one obtained from* B. xylophilus *genomic DNA.*

satellite DNA family as the target sequence for amplification should confer the additional advantage of specificity to the method. Moreover, due to the highly repetitive nature of satellite DNA, a greater sensitivity was obtained than with PCR protocols using single- or low-copy target sequences, and therefore this method allows a reliable detection from a single nematode.

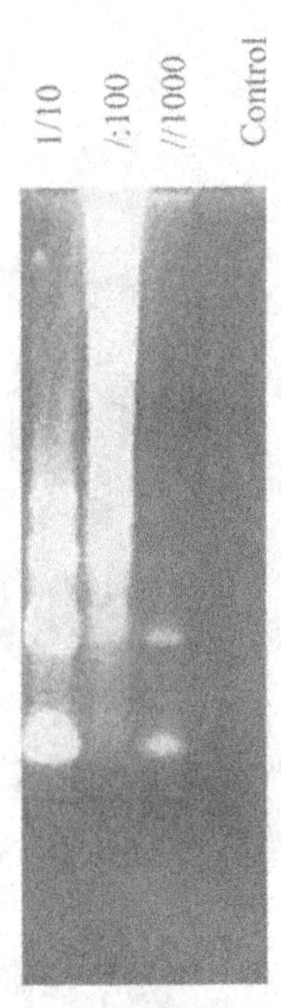

Fig. 3. *PCR experiments showing detection of* Bursaphelenchus xylophilus *in a mixture of other nematodes. The presence of one individual of* B. xylophilus *can be detected in mixture with 1000* B. mucronatus *nematodes.*

With primers deduced from the sequence of the monomeric unit we selected the optimal PCR conditions on genomic DNA from *Bursaphelenchus.*

To test the sensitivity of this method, amplification of DNA from single nematodes was considered. It was therefore necessary to select a simple method that could reliably make the DNA of a single individual suitable as a template for PCR. The use of proteinase K, in combination with alternating high and low temperatures, proved effective. PCR products from *B. xylophilus* single adults obtained by this method were thus compared to those obtained from *B. xylophilus* phenol/chloroform-purified genomic DNA.

As shown in Fig. 2, individual amplification patterns were always very similar to the one obtained from *B. xylophilus* genomic DNA. In a very few cases, little or no amplified product was obtained, presumably due to insufficient template (line 1 of individual PCR amplification for J10 strain). As expected, amplification was detected only in lanes corresponding to *B. xylophilus* isolates, and a regular ladder

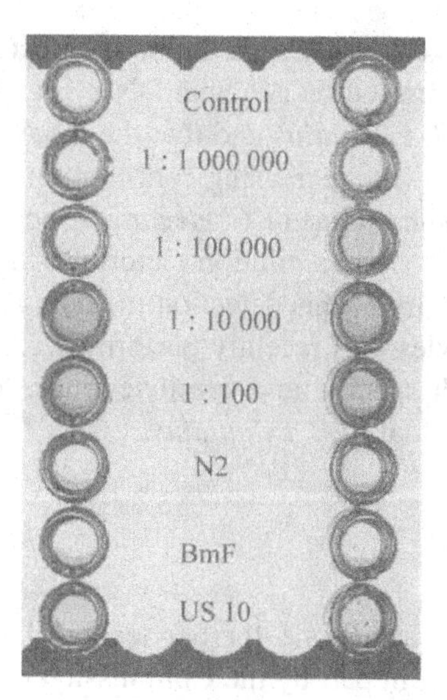

Fig. 4. *PCR amplification followed by automated UV-light detection. The amplified products were visualised by addition of an internal probe of satellite DNA labelled with DIG and a peroxidase-labelled antibody directed against DIG. A positive signal for* Bursaphelenchus xylophilus *was obtained up to a dilution factor of* 1 : 10^5 *with DNA of other species of nematodes such as* B. mucronatus *or* Caenorhabditis elegans.

pattern was obtained. In contrast, no amplification occurred in samples belonging to *B. mucronatus* prepared in the same manner. We defined the experimental conditions of the PCR procedure which allowed the specific amplification of *B. xylophilus* genomic sequences in a mixture of other nematodes such as *B. mucronatus* and *Caenorhabditis elegans.* The presence of one individual of *B. xylophilus* can be detected in mixture with 1000 *B. mucronatus* nematodes (Fig. 3).

PCR amplification followed by automated UV-light detection

In order to circumvent the labour intensive and time consuming electrophoretic separation of specific PCR amplification products, we developed a simple method using immunoenzymatic detection of the PCR product (PCR-ELISA) which allows automated detection of the *B. xylophilus* nematodes.

The specific target DNA is amplified using biotin-labelled primers and transferred to microplates coated with streptavidin. The labelled PCR products are then visualised by addition of an internal probe labelled with DIG and a peroxidase-labelled antibody directed against DIG. This method was first optimised on genomic DNA. Fig. 4 shows that a signal was obtained with *B. xylophilus* and that the DNA of *B. xylophilus* was detected when mixed with increasing concentrations of DNA from other species such as *B. mucronatus* or *C. elegans*. A positive signal for *B. xylophilus* was obtained up to a dilution factor of $1:10^5$.

To test the sensitivity of this detection method, amplification of DNA from single nematodes was recently performed. This detection method appeared to be both specific and sensitive since detection is observed with only one individual of *B. xylophilus*.

Acknowledgements

I thank Chantal Castagnone for her technical asssistance. This research was supported in part by the Commission of the European Community (Grant FAIR1-CT95-0034) which is gratefully acknowledged.

References

DE GUIRAN, G. & BRUGUIER, N. (1989). Hybridization and phylogeny of the pine wood nematode (*Bursaphelenchus* spp.). *Nematologica* 35, 321-330.

DE GUIRAN, G., LEE, M.J., DALMASSO, A. & BONGIOVANNI, M. (1985). Preliminary attempt to differentiate pinewood nematodes (*Bursaphelenchus* spp.) by enzyme electrophoresis. *Revue de Nématologie* 8, 85-92.

HARMEY, J. & HARMEY, M. (1993). Detection and identification of *Bursaphelenchus* species with DNA fingerprinting and polymerase chain reaction. *Journal of Nematology* 25, 406-416.

KIYOHARA, T. & BOLLA, R. (1990). Pathogenic variability among populations of the pinewood nematode *Bursaphelenchus xylophilus*. *Forest Science* 36, 1061-1076.

MAMIYA, Y. & ENDA, N. (1979). *Bursaphelenchus mucronatus* n. sp. (Nematoda: Aphelenchoididae) from pine wood and its biology and pathogenicity to pine tree. *Nematologica* 25, 353-361.

TARÈS, S., ABAD, P., BRUGUIER, N. & DE GUIRAN, G. (1992). Identification and evidence for relationships among geographical isolates of *Bursaphelenchus* spp. (Pinewood nematode) using homologous DNA probes. *Heredity* 68, 157-164.

TARÈS, S., LEMONTEY, J.M., DE GUIRAN, G. & ABAD, P. (1993). Cloning and characterization of a highly conserved satellite DNA sequence specific for the phytoparasitic nematode *Bursaphelenchus xylophilus*. *Gene* 129, 269-273.

TARÈS, S., LEMONTEY, J.M., DE GUIRAN, G. & ABAD, P. (1994). Use of species-specific satellite DNA from *Bursaphelenchus xylophilus* as a diagnostic probe. *Phytopathology* 84, 294-298.

WEBSTER, J.M., ANDERSON, R.V., BAILLIE, D.L., BECKENBACH, K., CURRAN, J. & RUTHERFORD, T.A. (1990). DNA probes for differentiating isolates of the pinewood nematode species complex. *Revue de Nématologie* 13, 255-263.

Nematology Monographs & Perspectives, 2003, Vol. 1, 165

PWN-CD: a taxonomic database for the pine wood nematode, *Bursaphelenchus xylophilus*, and other *Bursaphelenchus* species

Paulo VIEIRA [1], Jon EISENBACK [2] and Manuel M. MOTA [1]

[1] *NemaLab/ICAM/University of Évora, 7000 Évora, Portugal*
[2] *Department of Plant Pathology and Weed Science, Virginia Tech., Blacksburg, VA 24061, USA*

Summary – The main objective of this project is to develop an information database with the inclusion of text (research articles, keys, bibliography, *etc.*), images (jpeg files), video clips (mpeg files), in order to collect in one physical format (CD-ROM) all relevant information related to the taxonomy of pine wood nematode (PWN), *Bursaphelenchus xylophilus*, as well as other *Bursaphelenchus* species. Initially, all major research papers will be scanned in 'pdf' format. All available images of quality and importance to the subject will also be scanned from either journals, bulletins, reports, or obtained privately from colleagues' image collections. In certain cases, image quality (brightness, contrast, colour) may be enhanced with Adobe Photoshop. Short video clips will be compressed to 'mpeg' format for inclusion in the CD-ROM. An updated and broad bibliography will be included, as well as an intuitive taxonomic key, based on classical dicotomous keys but with a better and more intuitive presentation. Pre-existing images will be used in most cases; however, nematode specimens may need to be observed and photographed (jpeg image) in the lab. All the incorporated elements will be interlinked for ease of search from text to image to video. The total collection of images may be viewed individually using 'Extensis Portfolio'. The final product will be PC and Mac compatible. Future editions/updates will be developed. Ultimately, researchers, technicians and political decision-makers will have at their disposal a practical and readily available source of information on this extremely important pest and pathogen. Colleagues working with PWN are encouraged to provide materials for this task.

Nematology Monographs & Perspectives, 2003, Vol. 1, 167-176

Some environmental conditions that influence the spread of pine wilt

Kazuyoshi FUTAI [1] and Tamio AKEMA [2]

[1] *Graduate School of Agriculture, Kyoto University, Kyoto, Japan*
[2] *Kyushu Research Center, Forestry and Forest Products Research Institute, Kyushu, Japan*

Summary – The present study was carried out to reveal how environmental factors influence pine wilt development, and how such factors make a difference in the severity of pine wilt damage between different localities. A field survey was carried out from 1993-1998 at an experimental station of the university forest in Yamaguchi prefecture, located at the westernmost point of mainland Japan. The growth of 4000 seedlings of 23 families of various provenances of Japanese black pine *Pinus thunbergii*, and Japanese red pine *Pinus densiflora* was studied on a pine wilt affected site. We conclude that: *i*) drought exacerbates pine wilt damage; *ii*) mycorrhizal association mitigates drought stress, thereby reducing pine wilt damage; *iii*) topographical variability in a stand leads to differences in drought stress and in mycorrhizal activity, thereby causing the difference in pine wilt damage between localities within a stand.

Since the discovery of the pine wood nematode, *Bursaphelenchus xylophilus*, as the causal agent of mass death of pine trees in Japan, environmental factors such as temperature, precipitation, *etc*., have been studied intensively. Such factors seem to be crucial elements affecting the epidemic development of this forest disease. In particular, temperature has been regarded as one of the decisive factors in pine wilt disease. Precipitation is another factor that may determine pine wilt spread, because wilting is a typical symptom, and drought has been thought to promote the disease. Pine species are well-known ectomycorrhizal plants, and have symbiotic relationships with various basidiomycetous fungi, thereby obtaining mineral nutrients and water. Therefore, whether or not the mycorrhizal relationship is suitably established may be another determinant of this epidemic disease. Pine wilt has been reported to infect many species of the genus *Pinus*. The intensity of the damage, however, is known to vary from species to species (Futai & Furuno, 1979), between provenances within a species, and from one place to another within a stand. The present study was carried out to reveal how environmental factors influence pine wilt development, and how such factors make a difference in the severity of pine wilt damage between different localities.

Materials and methods

EXPERIMENTAL SITE

The field survey was carried out from 1993-1998 at an experimental station of the university forest in Yamaguchi prefecture, located at the westernmost point of mainland Japan. To compare growth between various provenances of Japanese black pine *Pinus thunbergii*, and Japanese red pine *P. densiflora*, *ca* 4000 pine seedlings of 23 families collected from various districts in Yamaguchi prefecture were planted on a slope on the station in 1973. The area of the stand is *ca* 1.4 ha, and the slope has an incline of 25°, and a height of *ca* 50 m.

Since 1979 pine wilt has spread into this stand and had killed *ca* 5% of the trees by the end of 1980. Until 1988, for 15 years after afforestation, the annual loss of pine trees due to pine wilt was rather low; only 23% of the total trees initially planted had been killed. By the end of 1993, the pine wilt damage had become more severe, and more than 70% of the pine trees had been killed (Fig. 1). At the end of 1993, however, a higher ratio of the provenances had survived. Hence we determined to follow the process of pine wilt spread into two provenances of Japanese black pine surviving at high ratio (no. 236, 35% survival; no. 241, 50%). The location of the provenances used for the present study (nos 236 and 241) is shown as a framed area in Fig. 1.

MYCORRHIZAL DENSITY AT DIFFERENT HEIGHTS OF A SLOPE

The mycorrhizal density was examined at four different heights of the slope. For each height ten soil samples were obtained. Each sample size was 30 × 30 × 15 cm. Fine roots less than 2 mm diam. were sorted out from each sample, divided into mycorrhizal and non-mycorrhizal, and their dry weights were measured separately. Mycorrhizal ratio was calculated by dividing the dry weight of mycorrhizal roots by that of total fine roots, and estimated as mycorrhizal density.

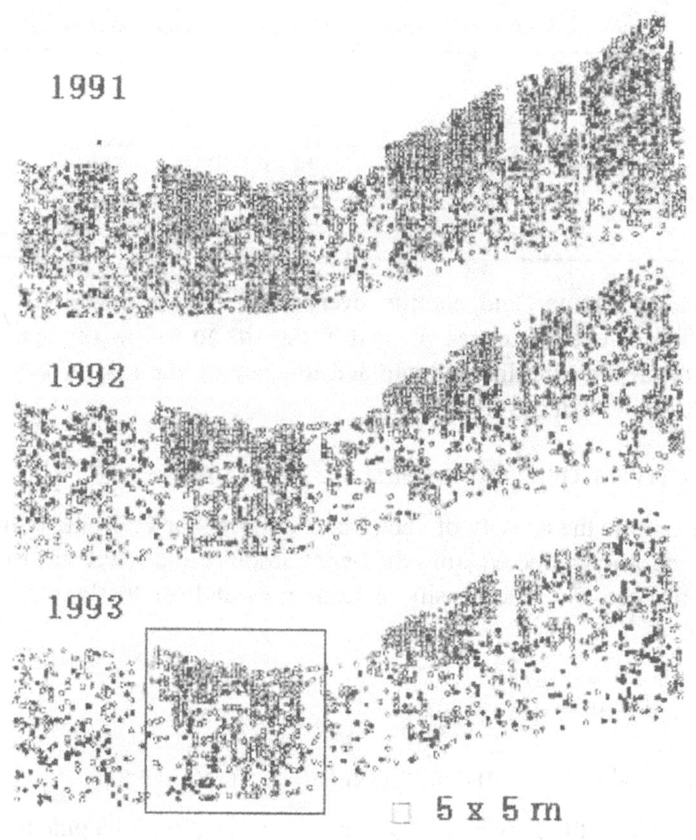

Fig. 1. *Decline of pine stand from 1991-1993. The squared area shows the location of provenances nos 236 and 241 which served for the present study.*

THE ROLE OF LATENT CARRIER TREES

Futai (1999) proposed the important role of the latent carrier trees that maintain a healthy appearance till the next summer though they have already ceased resin exudation. The hypothesis is as follows: such latent carrier trees were infected in the previous year, but had not yet shown wilting symptoms. When it became warmer, pathogenic nematodes surviving in healthy-looking trees began to increase their population and the host tree became diseased. When such latent carrier trees begin to show symptoms, they release gaseous substances such as alcohol and terpenes, thereby attracting vector beetles. If such beetles carry pathogenic nematodes, the surrounding trees attacked by the vector beetles will also be diseased.

Table 1. *Survival ratio (%) of pine trees in two provenances on a slope.*

Position on slope	Provenance	
	241	236
Upper	74	65
Middle	55	39
Lower	35	17

In early summer and autumn every year, we examined the resin exudation of each pine tree planted at the site to follow the spread of pine wilt at the experimental site and to examine the role of the latent carrier in pine wilt development.

The activity of vector beetle

To examine the activity of vector beetle, 35, 44 and 50 pine branches were randomly collected from the upper, middle, and lower parts of the slope, respectively. The density of feeding wounds on the branches was examined.

Results and discussion

Survival ratio at different heights of a slope

At the start of the study in 1993, the survival ratio of the pine trees on the slope was compared between different heights of the slope. As shown in Table 1, the survival ratio of two provenances of *Pinus thunbergii* was apparently highest at the upper part, followed by the middle part, and lowest at the lower part of the slope. Irrespective of the provenance of pines, survival ratio was highest at the upper part of the slope. This distribution of dead trees seemed very curious because rainwater always flows down the slope together with mineral nutrients.

Generally, therefore, the amount of mineral nutrients and water is less at the upper part of a slope than at the lower part. Water stress, therefore, seems more severe at the upper part than the lower part of a slope. Accordingly, pine wilt damage is supposed to be more severe on the upper part of a slope. Our findings, however, proved the contrary. A possible explanation for this curious phenomenon is as follows. Pine trees are well adapted to such a locality as the upper part of a slope is where most other plants cannot grow well and/or cannot survive.

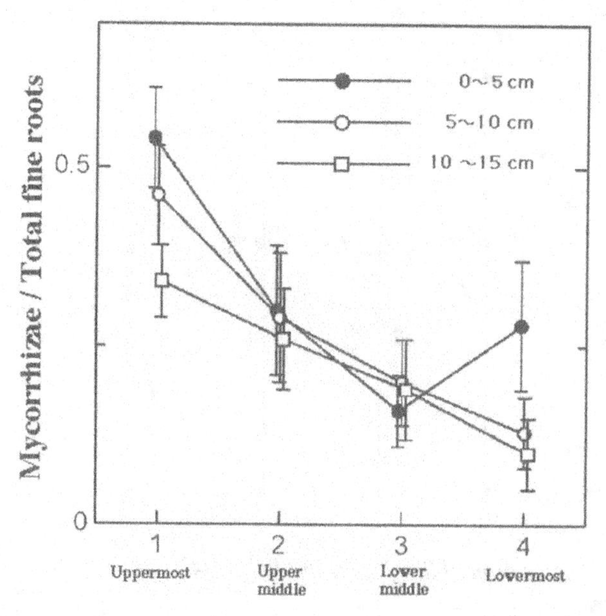

Fig. 2. *Mycorrhizal density at three depths in samples from four different positions on a slope.*

Species of the genus *Pinus* are well-known ectomycorrhizal plants and obtain part of their mineral nutrients and water from fungal symbionts; the fungi obtain photosynthate from pine trees in return. Under poor nutrient and water conditions, therefore, mycorrhizal relationships may develop to mitigate nutrient deficiency and reduce drought stress. At the lower part of a slope, however, there are sufficient nutrients and water to nourish pine trees without compensation by mycorrhizal association. When severe drought strikes the slope, however, the pine trees on the upper slope could survive because of mycorrhizal mitigation, while those on the lower part would face fatal stress. Thus, the survival ratio of pine trees on a slope may be higher on the upper part than the lower.

Mycorrhizal density at different heights of a slope

As shown in Fig. 2, mycorrhizal ratios were higher on the upper slope than on the lower part. Under normal climatic conditions, therefore, all pine trees planted on a slope are physiologically equal to each other. That is, pine trees on the lower part of the slope can absorb nutrients and water from surrounding soils where enough of these substances are present.

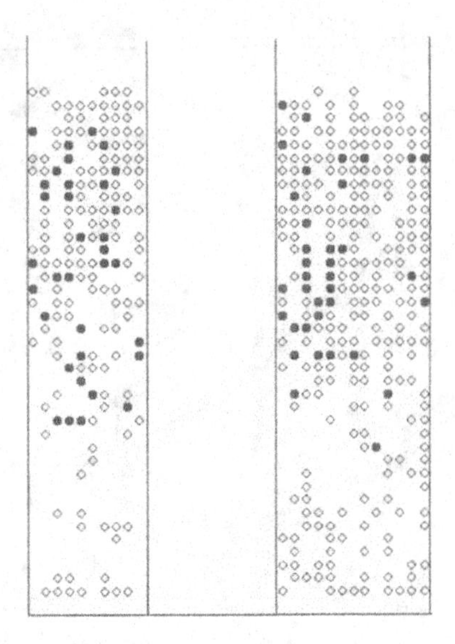

Fig. 3. *The distribution of dead trees in May 1994.*

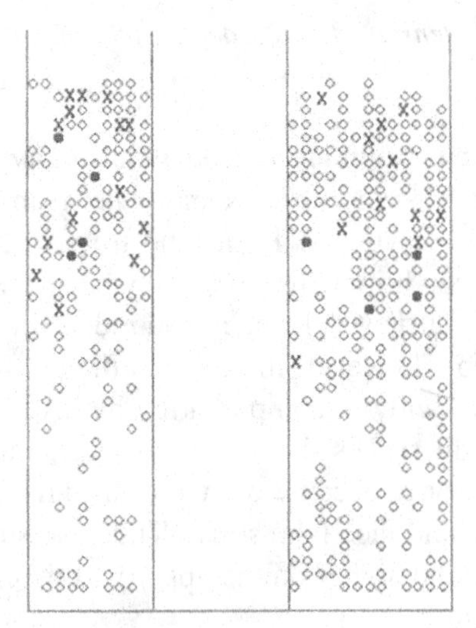

Fig. 4. *The distribution of newly-killed pine trees (shown as crosses) and that of latent carrier trees (shown as closed circles), October 1994.*

On the other hand, pine trees on the upper part of the slope are always under stress from drought and nutrient deficiency. These plants, however, mitigate their stress by mycorrhizal symbionts, thereby reducing drought stress and nutrient deficiency.

THE ROLE OF LATENT CARRIER TREES

If pine trees planted at any height of the slope are equally vulnerable/resistant to pine wilt infection, the number of dead trees should be proportional to the number of surviving trees irrespective of the height on the slope. As shown in Fig. 3, dead trees in 1993 and 1994 were not distributed as we had supposed. The trees shown as closed circles in this figure were killed by May 1994 due to the pine wilt infection in the previous year. After dead trees were removed in early spring of 1994, we drilled a tiny hole into the bark (5 mm in diam.) of all the remaining trees to examine whether they could exude resin or not. As shown in Fig. 4, we found eight 'latent carrier trees', shown as closed circles. In October 1994, newly killed trees (shown as crosses) appeared around the latent carrier trees, which also had been killed by this time. Thus, the trees newly killed in 1994 seemed to occur in the vicinity of the latent carrier trees.

In November 1995, we examined the latent infection by *B. xylophilus* by the following method; many pine branches were randomly collected from healthy-looking pine trees, one branch from each tree. Among them, 20 branches with feeding wounds of *M. alternatus* served for nematode extraction. *Bursaphelenchus xylophilus* nematodes were found in five of 20 branches examined. Thus, pine wood nematodes were confirmed to survive in the healthy-looking trees.

In 1995, we surveyed the distribution of dead pine trees twice, in June and November. Dead trees in June were so-called latent carrier trees which had been infected in the previous year, survived the winter, and became diseased in June 1995. The distribution of the dead trees expanded to the lower part of the slope. In November, newly-killed trees and diseased trees, which ceased their resin exudation, occurred over the slope, especially around the lower part of the slope.

THE ACTIVITY OF VECTOR BEETLES

The spread of pine wilt into the lower part of the slope in 1995 could be attributed to the biased activity of the vector beetle, *Monochamus alternatus*. Between three heights of the slope there was no critical difference in frequency distribution of feeding wound density. This suggests no critical difference in activity of the vector beetle, *M. alternatus* between different heights of the slope.

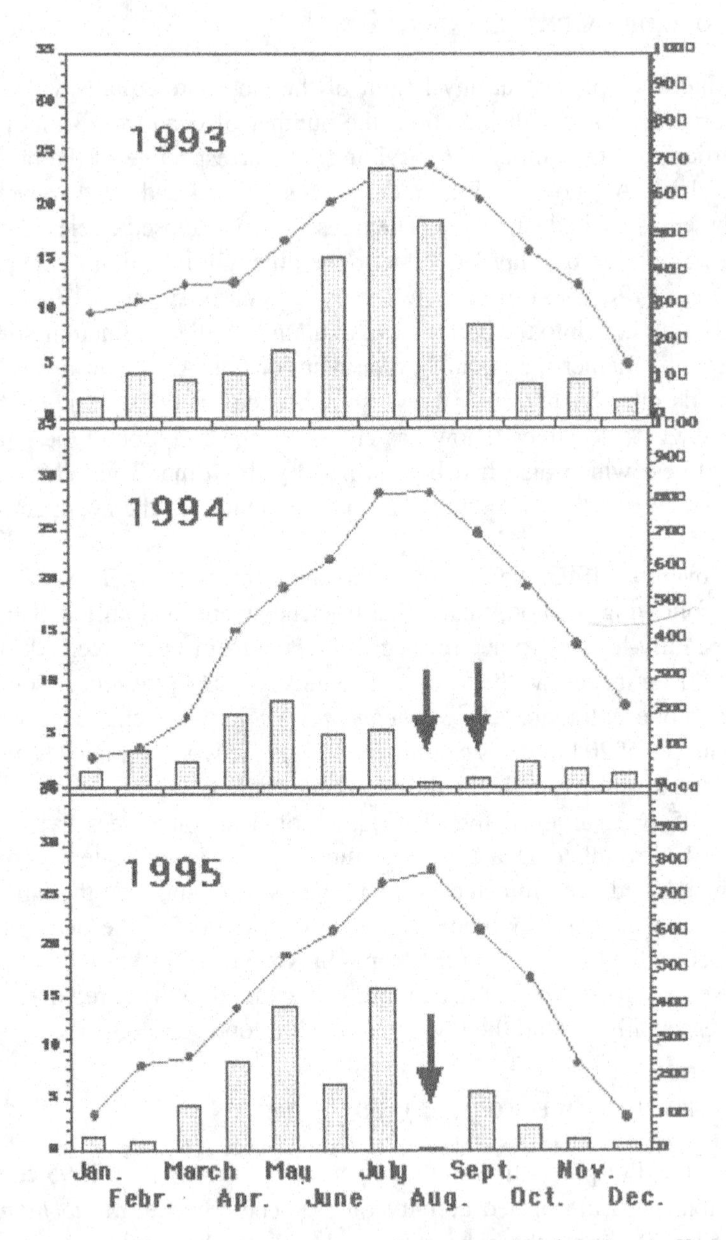

Fig. 5. *The annual climatic changes at the experimental station in Yamaguchi prefecture from 1993-1998. Bars and dotted lines show the precipitation and temperature, respectively.*

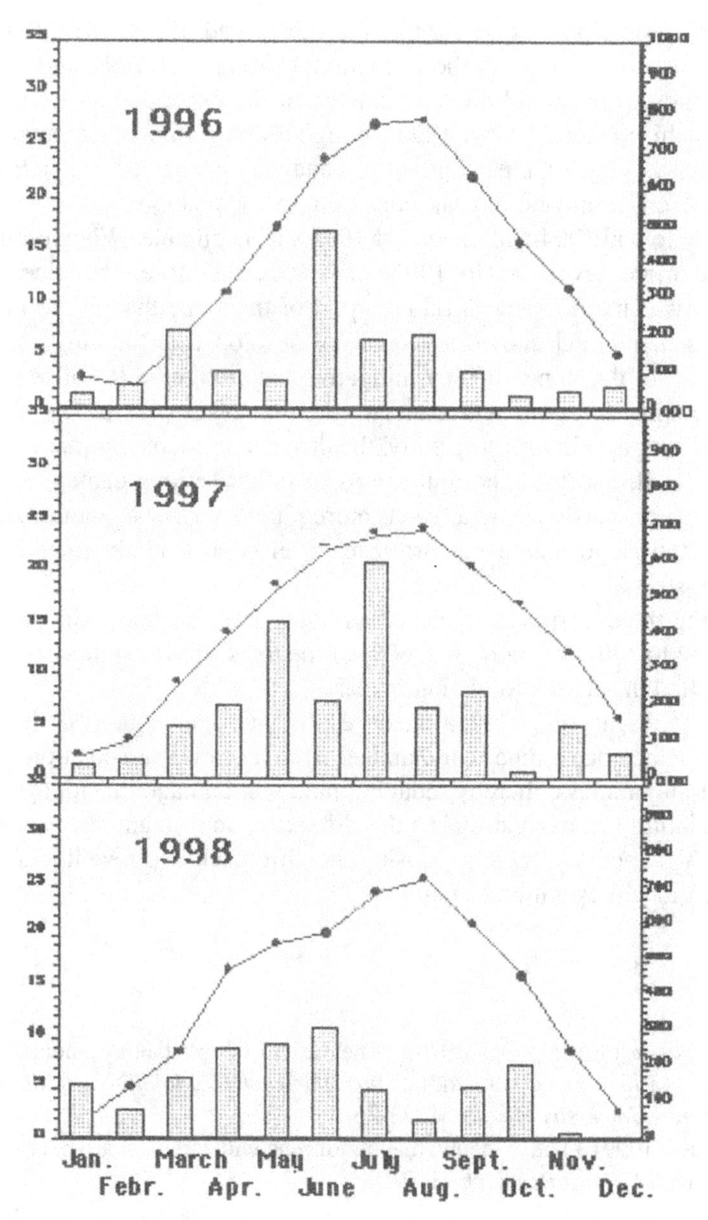

Fig. 5. *(Continued).*

DROUGHT AND SEVERITY OF PINE WILT

Another possible reason for the unequal spread of pine wilt into the lower part of the slope is the exceptional drought of 1994 and 1995. Fig. 5 shows the annual climatic change in the experimental station at Yamaguchi prefecture from 1993 through 1998. Summer temperatures were always very high, peaking in July and August, but the precipitation was different from year to year. For example, the precipitation in August was very low in 1994 and in August 1995 was negligible. When drought became more severe as in 1994 and 1995, therefore, the pine wilt damage was more severe at a lower part of the slope than at the upper part. As mentioned above, mycorrhizae develop conspicuously at the upper part of the slope, thereby mitigating the drought stress of summer and reducing the pine wilt damage. At the lower part of the slope, mycorrhizae develop insufficiently, the drought stress of summer cannot be mitigated, and the pine wilt seems to inflict serious damage at the lower part of the slope. As a result, more pine trees survive at the upper part of the slope where mycorrhizae develop sufficiently to mitigate drought stress.

During the experimental period, surviving trees decreased in number from 452 to 170, *i.e.*, more than 60% of the trees surviving in May 1993 were killed during the following 5 years.

Based on the above-mentioned results, we can conclude that: *i*) drought exacerbates pine wilt damage; *ii*) mycorrhizal association mitigates drought stress, thereby reducing pine wilt damage; *iii*) topographical variability in a stand makes the difference in drought stress and in mycorrhizal activity, thereby causing the difference in pine wilt damage between localities within a stand.

References

FURAI, K. & FURUNO, T. (1979). [The variety of resistances among pine-species to pine wood nematodes, *Bursaphelenchus xylophilus*.] *Bulletin of the Kyoto University Forests* 51, 23-36.

FUTAI, K. (1999). [The epidemic manner of pine wilt spread in a Japanese red pine stand.] *Forest Research* 71, 9-18.

Nematology Monographs & Perspectives, 2003, Vol. 1, 177-185

Persistence of the pine wood nematode in asymptomatic Scots pines

Dale R. BERGDAHL and Shari HALIK

Department of Forestry, University of Vermont, Burlington, 05405 Vermont, USA

Summary – In 1987 and 1993, 110 and 80 Scots pine trees, respectively, were inoculated with the pine wood nematode (PWN) (*Bursaphelenchus xylophilus*) and, in those years, an additional 30 and 80 trees, respectively, were inoculated with sterile distilled water to serve as controls. All trees were observed and sampled to determine how long PWN would persist in living trees; however, since 1998 only dead trees have been sampled for PWN. As we previously reported, these Scots pine trees were found to harbour populations of PWN for up to 11 years after inoculation without inciting pine wilt disease (PWD). However, some mortality has occurred in both PWN-inoculated and control trees. Pine wood nematodes were never recovered from control trees. For trees inoculated in 1993, over 50% were found to have a population of PWN by 1998. Most PWN-infested trees appeared healthy with live crown ratios ranging from 30-60%. Between 1998 and 2001, eight PWN-inoculated and 12 control trees died, but only inoculated trees contained PWN. Final sampling is scheduled for summers 2002 and 2003. The PWN can persist for many years in asymptomatic, living trees without inciting PWD which may hamper sanitation and other control efforts in the management of PWD. Thus, harvesting only healthy-appearing trees may not be an adequate measure to prevent movement of the PWN in roundwood or other wood products.

Bursaphelenchus xylophilus (Steiner & Buhrer) Nickle, the pine wood nematode (PWN), is the causal agent of pine wilt disease (PWD). This nematode is native to North America (NA) where it was first described in longleaf pine (*Pinus palustris* Mill.) but it was not known to be pathogenic (Steiner & Buhrer, 1934). However, this nematode is now considered one of the most important exotic forest pests creating havoc and concern around the world, especially in Asia (Nickle, 1985; Bergdahl, 1988; Liebhold *et al.*, 1995). Recently, PWN was found in the forested regions of Portugal on maritime pine (*P. pinaster* Aiton) which is the first report of this nematode in Europe (Mota *et al.*, 1999).

The pines (*Pinus* spp.) of NA are considered resistant to PWN while certain exotic species (*P. sylvestris* L., Scots pine; *P. nigra* Arnold, Austrian pine; and *P. thunbergii* Parlatore, Japanese black pine) growing there are considered susceptible and may develop PWD, especially when growing on certain sites (Wingfield *et al.*, 1982). The natural reservoir for the PWN in NA is recently killed trees that may also harbour a population of wood-staining and other fungi that may serve as a food source for PWN (Steiner & Buhrer, 1934). This nematode is commonly found in cut timber (especially logs) or associated with conifer trees dying from other biotic causes (Dwinell & Barrows-Broaddus, 1983; Wingfield & Blanchette, 1983; Bergdahl *et al.*, 1984; Bergdahl & Halik, 1986). In addition, abiotic stress factors, such as high temperatures and low soil moisture, are reported to predispose trees to infection by PWN (Wingfield *et al.*, 1982; Rutherford & Webster, 1987).

The PWN is vectored by long-horned beetles (*Monochamus* spp.) which are normal inhabitants of coniferous forests around the world. These vectors transmit the PWN to new host trees through wounds created during maturation feeding and/or during oviposition (Linit, 1988). *Monochamus alternatus* Hope and *M. carolinensis* (Olivier) are considered to be the most important vectors in Asia and NA, respectively (Linit, 1988), and *M. galloprovincialis* (Olivier) has been found infested with PWN in Europe (Mota, pers. comm.).

Pine wood nematode is known to persist in trees for extended periods of time following inoculation without causing symptoms of PWD, especially in cooler climates. In NA, PWN is reported in seemingly healthy Scots pines for up to 11 years following inoculation (Bergdahl & Halik, 1998). This nematode has also been found in asymptomatic trees in cooler areas of Japan but its significance relative to PWD is unknown (Futai, 1995). The main purpose of this paper is to report findings from our continued study on the long-term persistence of PWN in living Scots pines.

Materials and methods

Since the main purpose of this paper is to extend our findings from several previous reports (Bergdahl & Halik, 1994; Halik & Bergdahl, 1994; Bergdahl & Halik, 1998), details of the materials and methods for inoculation and sampling can be found in those papers; however, a brief summary is included below.

During the growing seasons of 1987 and 1993, 110 and 80 healthy, 20-year-old Scots pine trees, respectively, were inoculated with a Scots pine isolate of PWN and in those years an additional 30 and 80 trees, respectively, were inoculated with sterile distilled water to serve as controls. To inoculate, a hole (0.6 × 3.0 cm) was drilled into the main stem at 1.5 m above the ground on the southeast and southwest-facing sides of each tree. Approximately 30 000 and 5000 nematodes/ml of water were added to each inoculation wound in 1987 and 1993, respectively. A similar amount of sterile water was used for the controls. All inoculated trees were assessed each summer to determine their health, diameter growth, and live crown ratios. Diameter growth was measured at 1.3 m above the ground and live crown ratios were calculated as a percentage of total tree height.

To sample standing trees for PWN, a gasoline-powered drill with a 2.5 cm Forstner bit was used to remove bark and wood shavings suitable for extraction of PWN. In each sample tree, three holes were drilled to a depth of 3-5 cm in the vicinity of the inoculation point but not closer than 5 cm. All bark and wood shavings per sample tree were collected in a sterile, sealable plastic bag and then incubated for about 30 days at 25°C. After incubation, 300 ml of sterile distilled water were added to each bag. Bags were then incubated for 48 h at 25°C, after which the water was drained from the bags and refrigerated until the extracted nematodes were counted. All bark and wood samples were oven-dried, weighed, and numbers of nematodes reported on a per g oven-dry-weight basis. Trees were sampled for PWN on an annual basis through 1998 after which only trees that had died since that time were sampled during the year of death.

Results

The following is a summary of our PWN sampling from 1987 to 1998 as previously reported (Bergdahl & Halik, 1998). For the 1987 inoculations, PWN was never found in any control tree during the 11-year study. However, PWN was recovered from 25 trees (23%) at some time during the study and 72% of these trees were living and had healthy-appearing crowns at the time of the first positive extraction.

Table 1. *Recent mortality of Scots pines inoculated with pine wood nematode (PWN) in 1993 in Wolcott, Vermont, USA. A total of 80 trees were inoculated with PWN and 80 trees were inoculated with sterile distilled water as controls. Trees were 25 years old at time of inoculation.*

	Number of Scots pines				
	1993-1998	1999	2000	2001[1)]	Total
Current year mortality	16	8	3	9	36
PWN-inoculated	11[2)]	4	1	3	19
PWN recovered[3)]	9	4	0	3	16 (84)
Controls	5[2)]	4	2	6[4)]	17
PWN recovered[3)]	0	0	0	0	0

[1)] Does not include PWN recovery data for 2001.
[2)] One tree died from stem breakage.
[3)] PWN recovered from trees after inoculation (%).
[4)] Three trees died from stem breakage.

The numbers of PWN extracted ranged from 1-20 000/g of dry wood sampled. The majority of these PWN-infected trees remained healthy and asymptomatic for 7-11 years while harbouring PWN.

For the 1993 inoculations, the PWN was recovered from 42 of 80 (52%) inoculated trees but not from any control trees as of 1998. A total of five control trees had died by 1998 and, of the 42 infected trees, 76% were still living and 75% of the living trees had healthy-appearing crowns with live crown ratios ranging from 30-60%. The numbers of nematodes extracted from these trees ranged from 1-38 000/g of dry wood.

Annual sampling of living trees for PWN was discontinued in 1998 because of the excessive amount of damage that was occurring near the point of inoculation on the main stem. All trees will be sampled for PWN for the last time during August 2002 and 2003 for trees inoculated in 1987 and 1993, respectively.

Mortality of both PWN-inoculated and control trees has occurred at various intervals during both the 1987 and 1993 studies; however, only the results of the 1993 inoculations are presented. Prior to 1999, a total of 11 PWN-inoculated and five control trees had died (Table 1). By 2001, a total of 19 PWN-inoculated trees had died along with 17 controls; however, three of the control trees that died in 2001 were the result of stem breakage at the point of stem sampling. Pine wood nematode

Table 2. *Persistence of pine wood nematode (PWN) in living Scots pines inoculated with PWN in 1993 in Wolcott, Vermont, USA. A total of 80 trees were inoculated with PWN and 80 trees were inoculated with sterile distilled water as controls. Trees were 25 years old at time of inoculation.*

	Number of Scots pines			
	1998	1999	2000	2001
Current year survival	144	136	133	124
PWN-inoculated	69	65	64	61
PWN recovered[1)]	32 (46)	27 (42)	27 (42)	24 (39)
Controls	75	71	69	63
PWN recovered[1)]	0	0	0	0

[1)] PWN recovered from trees any year after inoculation (%).

was never recovered from any of the dead, control trees but had been previously recovered from 16 (84%) of the 19 dead PWN-inoculated trees.

The PWN is known to persist in living trees without causing PWD. For the 1993 inoculated trees, the PWN had been recovered, at least once, from 52% of these trees prior to 1998 when this percentage had dropped to 46% and by 2001 to 39% (Table 2). This drop in PWN recovery in living trees is primarily due to the increased mortality levels with time.

Both diameter growth and live crown ratios have been measured for all living, PWN and control inoculated trees since 1993. In general, diameter has increased slightly over the past 8 years for both PWN and control inoculated trees (Fig. 1). However, live crown ratios increased slightly between 1993 and 1995 for both PWN and control inoculated trees and then have since been on the decline to their current level of about 35% (Fig. 2).

Discussion

Scots pine is not native to North America but has been planted extensively for a variety of reasons. In many instances, the species has been planted on sites not well suited for growth and development, so trees often experience stress in the form of unusually high temperatures, low soil moisture regimes, and/or reduced nutrient levels. Under these conditions, Scots pines are known to be very susceptible to infection

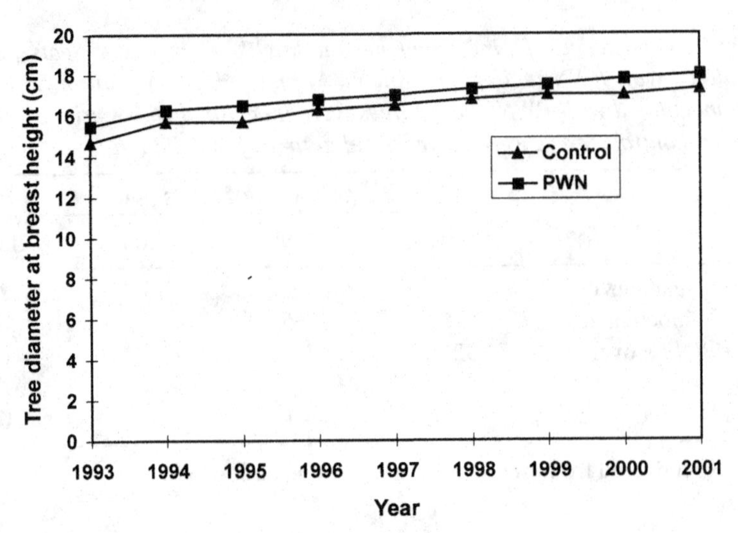

Fig. 1. *Average annual diameter measurements of surviving Scots pines inoculated with pine wood nematode (PWN) in 1993 in Wolcott, Vermont, USA.*

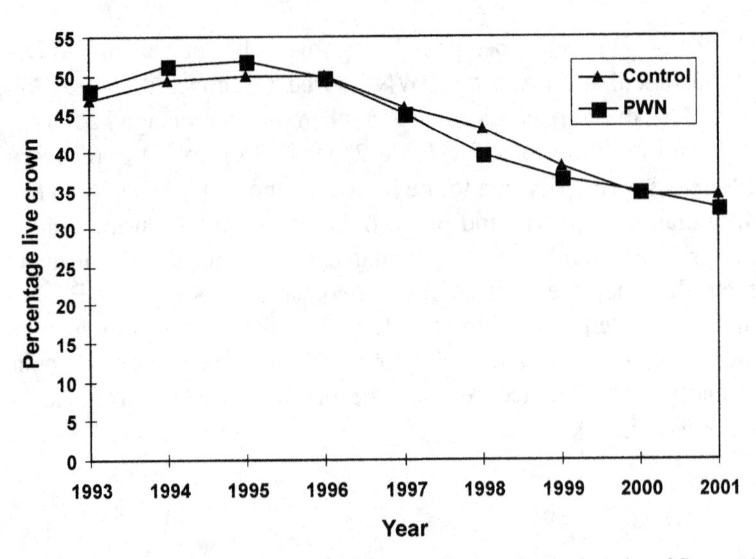

Fig. 2. *Average annual percentage of live crown/total tree height of Scots pines inoculated with pine wood nematode (PWN) in 1993 in Wolcott, Vermont, USA.*

by PWN and they rapidly develop symptoms of PWD (Robbins, 1982; Wingfield *et al.*, 1982).

The Scots pines in this study, however, are growing in a cool, northern climate in Vermont and the inoculated trees have served as long-term hosts for PWN as previously described (Halik & Bergdahl, 1994; Bergdahl & Halik, 1998). Some tree mortality has occurred, but the rapid wilt symptoms normally associated with PWD have not developed. Instead, the majority of these inoculated trees still maintain a small, viable population of PWN but has remained reasonably healthy for many years without developing symptoms of PWD. This lack of symptom development may be related to the high temperature thresholds described by Rutherford and Webster (1987) as being required for rapid increase in PWN populations and for wilt symptoms to occur. However, it is not known how temperatures influence the maintenance and persistence of PWN populations in asymptomatic trees.

Strict sanitation measures and the destruction of diseased trees are important components in the management of PWD. However, Futai (1995) recently raised concerns about the role of symptomless trees in maintaining the PWD epidemic in certain areas of Japan where it has been noted that removal of all symptomatic trees in one year may reduce, but does not eliminate, the need for removal of diseased trees the following year. It is not known if these newly diseased trees are the result of new vector transmissions or if they are simply the result of PWN residing in asymptomatic trees until conditions become suitable for development of PWD. In our opinion, the mere fact that PWN can persist in living trees creates a serious management problem for programmes dependent on sanitation procedures to limit dissemination of PWN and continued development of PWD. In addition, it should be noted that PWN is difficult to detect in wood, and that the harvesting of healthy-appearing trees is no guarantee they are free of the PWN. Therefore, the harvesting only of trees of healthy appearance may not be an adequate measure to prevent movement of the PWN in roundwood or other wood products.

References

BERGDAHL, D.R. (1988). Impact of pinewood nematode in North America: present and future. *Journal of Nematology* 20, 260-265.

BERGDAHL, D.R. & HALIK, S. (1986). *Bursaphelenchus xylophilus*-associated conifer mortality in the Northeastern United States. *American Phytopathological Society Symposium Proceedings: Biology and ecology of the pine wood nematode, Reno, Nevada, 1985*, St Paul, MN, USA, APS Press, pp. 46-49.

BERGDAHL, D.R. & HALIK, S. (1994). Persistence of *Bursaphelenchus xylophilus* in living *Pinus sylvestris. Phytopathology* 83, 242.

BERGDAHL, D.R. & HALIK, S. (1998). Inoculated *Pinus sylvestris* serve as long-term hosts for *Bursaphelenchus xylophilus.* In: Futai, K., Togashi, K. & Ikeda, T. (Eds). *Sustainability of pine forests in relation to pine wilt and decline. Proceedings of the Symposium, Tokyo, Japan, 26-30 October 1998.* Kyoto, Japan, Shokado Shoten, pp. 73-78.

BERGDAHL, D.R., SMELTZER, D.L.K. & HALIK, S.S. (1984). Components of conifer wilt disease complex in the northeastern United States. *Proceedings of the Joint US/Japanese Pine Wilt Disease Seminar, Honolulu, Hawaii*, pp. 152-157.

DWINELL, L.D. & BARROWS-BROADDUS, J. (1983). Pine wilt and pitch canker of Virginia pines in seed orchards. *Proceedings of the 17th Southern Tree Improvement Conference, Athens, GA, USA*, pp. 55-62.

FUTAI, K. (1995). The role of the symptomless carrier in epidemic spread of the pine wilt disease. *Proceedings of the international symposium on pine wilt disease caused by the pine wood nematode.* Beijing, China, Chinese Society of Forestry, pp. 69-80.

HALIK, S. & BERGDAHL, D.R. (1994). Long-term survival of *Bursaphelenchus xylophilus* in living *Pinus sylvestris. European Journal of Forest Pathology* 24, 357-363.

LIEBHOLD, A.M., MACDONALD, W.L., BERGDAHL, D.R. & MASTRO, V.C. (1995). Invasion by exotic forest pests: a threat to forest ecosystems. *Forest Science Monograph* 30, 1-49.

LINIT, M.J. (1988). Nematode-vector relationships in the pine wilt disease system. *Journal of Nematology* 20, 227-235.

MAMIYA, Y. (1988). History of pine wilt in Japan. *Journal of Nematology* 20, 219-226.

MOTA, M.M., BRAASCH, H., BRAVO, M.A., PENAS, A.C., BURGERMEISTER, W. METGE, K. & SOUSA, E. (1999). First report of *Bursaphelenchus xylophilus* in Portugal and Europe. *Nematology* 1, 727-734.

NICKLE, W.R. (1985). Pine wood nematode causing raw wood export problems. *Journal of Nematology* 17, 506.

ROBBINS, K. (1982). Distribution of the pinewood nematode in the United States. In: Appleby, J.E. & Malek, R.B. (Eds). *Proceedings of the National Pine Wilt Disease Workshop, Urbana, IL, USA*, Illinois Natural History Survey, pp. 3-6.

RUTHERFORD, T.A. & WEBSTER, J.M. (1987). Distribution of pine wilt disease with respect to temperature in North America, Japan and Europe. *Canadian Journal of Forest Research* 17, 1050-1059.

STEINER, G. & BUHRER, E.M. (1934). *Aphelenchoides xylophilus*, n. sp., a nematode associated with blue-stain and other fungi in timber. *Journal of Agricultural Research* 48, 949-951.

TOMMINEN, J. (1991). Pinewood nematode, *Bursaphelenchus xylophilus*, found in packing case wood. *Silva Fennica* 25, 109-111.

WINGFIELD, M.J. & BLANCHETTE, R.A. (1983). The pinewood nematode, *Bursaphelenchus xylophilus*, in Minnesota and Wisconsin: insect associates and transmission studies. *Canadian Journal of Forest Research* 13, 1068-1076.

WINGFIELD, M.J., BLANCHETTE, R.A., NICHOLLS, T.H. & ROBBINS, K. (1982). Association of pinewood nematode with stressed trees in Minnesota, Iowa and Wisconsin. *Plant Disease* 66, 934-937.

Nematology Monographs & Perspectives, 2003, Vol. 1, 187-197

Quarantine concerns about the methods used to demonstrate pathogenicity of *Bursaphelenchus* spp.

David G. McNAMARA

EPPO, 1 rue Le Nôtre, 75116 Paris, France

Summary – It is clear that experiments in which *Bursaphelenchus* spp. are inoculated into young plants have no obvious scientific merit and they provide no relevant information about pathogenicity. Furthermore, the finding of a particular nematode species in a dead tree does not prove that the nematode was responsible for the death of the tree. Therefore, it must be concluded that the publications on the pathogenicity of *Bursaphelenchus* spp., other than *B. xylophilus*, do not present adequate evidence to support the view that these species could adversely affect trees of pine under natural conditions. *Bursaphelenchus mucronatus, B. sexdentati, B. leoni, B. hellenicus* or *B. hofmanni* cannot be considered to be pathogenic to *Pinus* spp. unless further convincing evidence is produced. *Bursaphelenchus xylophilus* remains the only species of the genus that is pathogenic to trees, and only to a limited number of *Pinus* spp. There is a need to develop methodologies to demonstrate and prove pathogenicity by *Bursaphelenchus* spp.

For the past 30 years, it has been known that *Bursaphelenchus xylophilus* can cause the death of healthy pine trees in the forests (Kiyohara & Tokushige, 1971). This pathogenicity occurs only on a limited number of non-American species of *Pinus* in North America and in other parts of the world where this nematode has been introduced. No other cases of similar tree mortality in the field by any of the other approximately 42 *Bursaphelenchus* spp. have been reported. However, several scientists have reported that the closely related *B. mucronatus* is pathogenic to young plants of *Pinus* following artificial inoculation of nematodes (Cheng *et al.*, 1986; Yang *et al.*, 1988; Schauer Blume, 1990; Bakke *et al.*, 1991; Riga *et al.*, 1991; Braasch, 1996; Braasch *et al.*, 1998). Furthermore, in recent years, several other *Bursaphelenchus* spp. have also been reported to kill young pine plants (Braasch *et al.*, 1998; Skarmoutsos & Michalopoulos-Skarmoutsos, 2000). Table 1 presents all

Table 1. Bursaphelenchus *species (apart from* B. xylophilus*) reported to kill* Pinus *spp.*

Nematode species	*Pinus* spp. killed
B. mucronatus	*P. sylvestris, P. pinaster, P. nigra, P. massoniana, P. thunbergii*
B. sexdentati	*P. sylvestris, P. pinaster, P. nigra*
B. leoni	*P. halepensis*
B. hellenicus	*P. pinaster, P. bruta*
B. hofmanni	*P. sylvestris*

these reported cases of *Bursaphelenchus* spp., other than *B. xylophilus*, which kill species of *Pinus*.

If these reports are to be believed, then it is possible that these other *Bursaphelenchus* spp. are responsible for the death of healthy trees in the forests, but that this fact has been overlooked and/or the cause of death has been ascribed to other causes. This possibility is of particular relevance to plant quarantine, since the nematodes concerned are of limited global distribution and could be transported to different parts of the world with the increasing trade in wood. The question is raised, therefore, as to whether plant quarantine services in different parts of the world should apply phytosanitary measures to try to prevent the spread of these *Bursaphelenchus* spp.

The purpose of this paper is to try to answer this question by examining the reports of pathogenicity of *Bursaphelenchus* spp. other than *B. xylophilus*, particularly in relation to the methods used to demonstrate pathogenicity, and to try to conclude whether the reports should be taken seriously. Although some of the reports mention, as supporting evidence, that nematodes have been recovered from dead trees in the field, the main evidence presented for pathogenicity is based on the results of experiments in which nematodes are inoculated into young plants.

The use of inoculation experiments in young plants has been criticised on several occasions by eminent authorities in the study of *Bursaphelenchus* (Mamiya, 1983; Wingfield & Blanchette, 1984; Bedker, 1987; Dwinell, 1997), on the grounds that the effects of nematodes on such young plants cannot be taken to represent the response of large trees. Nevertheless, research workers continue to use such methods, sometimes arguing that it is the only method open to them because inoculation of

Table 2. *Some examples of variability of results in inoculation experiments with young plants of* Pinus *spp. Figures represent the percentage of plants killed by* Bursaphelenchus xylophilus *in different experiments on the same host species.*

P. banksiana	0	0	19	25	75	87	
P. resinosa	0	5	13	21	30	36	88
P. strobus	0	3	10	44	75	100	
P. massoniana	0	10	"moderate"		90		
P. taeda	0	7	25	47			

Includes data from: Bai and Cheng (1993), Burns *et al.* (1985), Dropkin *et al.* (1981), Dwinell (1985), Myers (1984), Ohba *et al.* (1984), Shauer-Blume (1990), Sutherland *et al.* (1991), Wingfield *et al.* (1983), Yang *et al.* (1987).

large trees in the field with *B. xylophilus* is difficult as well as being dangerous from a phytosanitary point of view, due to the risk of spread of this serious pest.

With the aim of evaluating the use of inoculation experiments on young plants, this paper presents a review of the available literature wherein such inoculation experiments for studying comparative pathogenicity rates of *B. xylophilus* are described.

LITERATURE REVIEW OF PATHOGENICITY METHODS

The inoculation experiments in this literature review were conducted on *B. xylophilus* and were generally of the same type as those on the species listed in Table 1. Young plants aged 1-5 years old were wounded either on the growing tip, the shoot or the main stem and a suspension of nematodes, containing between several tens of nematodes and several thousands, was applied to the wound. Inoculation experiments of this type have been used to study various aspects of host parasite relationships, such as histology, physiology, control, *etc.*, but only those whose aim was comparative pathogenicity between different *Pinus* spp. or between different nematode populations, and in which the conclusion of the experiment was the death or survival of the plants, were included in this review.

The first point that becomes obvious is the great variability of the results (Table 2). Of course, such variability may be due to differences in populations of nematodes, numbers of nematodes inoculated, life stage of the nematodes, provenance of the seedlings, health status of the seedlings, environmental conditions and inoculation techniques; the

Table 3. Pinus *species killed by* Bursaphelenchus xylophilus *in seedling inoculation tests. Species arranged in order of mean 'susceptibility'* i.e., *the average proportion of plants killed in the different experiments; Number 1 is the most 'susceptible' and number 34 the least. The species underlined are the only species known to be killed by natural infection of* B. xylophilus *in the field.*

1. *caribaea*	10. *radiata*	19. *banksiana*	28. *nigra*
2. *pseudostrobus*	11. *luchuensis*	20. *yunanensis*	29. *fenzeliana*
3. *oocarpa*	12. *flexilis*	21. *kwangtungensis*	30. *contorta*
4. *echinata*	13. *elliottii*	22. *koraiensis*	31. *ponderosa*
5. *thunbergii*	14. *densiflora*	23. *bungeana*	32. *taeda*
6. *morrisonicola*	15. *strobiformis*	24. *massoniana*	33. *armandii*
7. *lambertiana*	16. *monticola*	25. *resinosa*	34. *taiwanensis*
8. *rigida*	17. *sarotina*	26. *virginiana*	
9. *sylvestris*	18. *strobus*	27. *griffithii*	

Includes data from: Bai and Cheng (1993), Bakke *et al.* (1991), Bedker *et al.* (1984), Braasch (1997), Burns *et al.* (1985), Dropkin *et al.* (1981), Dwinell (1984, 1985), Kaneko and Zinno (1986), Lee (1986), Mamiya and Kiyohara (1972), Myers (1984), Ohba *et al.* (1984), Shauer-Blume (1990), Sutherland *et al.* (1991), Tamura and Dropkin (1984), Wingfield *et al.* (1983), Yang *et al.* (1988).

influence of some of these factors has been demonstrated by Chang and Chang (1999) and Kishi (1999).

A much more serious criticism, however, is the very obvious fact that there is no apparent relationship between the results of inoculation experiments on young plants and the occurrence of the disease in the field. All of the *Pinus* species listed in Table 3 have been tested in inoculation experiments on young plants and have been killed by *B. xylophilus*. The fact is that, in the field, only seven species of *Pinus* are known to be killed by *B. xylophilus* (the six species underlined in Table 3, and *P.* pinaster). Furthermore, those species that are really killed by *B. xylophilus* in the field do not appear at the top of the list of susceptibility (as arranged in Table 3, based on the average proportion of killed plants from all the different experiments). Therefore, if the reader did not know the field situation, the results of these experiments would give absolutely no indication of it.

The lowest mortality in the list in Table 4 is approximately 25%, which appears to be quite a high level of pathogenicity (even though some authors have classified it as 'moderate' or even 'low'); it could

Table 4. *Plants of* Pinus *spp. older than 5 years killed by inoculation with* Bursaphelenchus xylophilus. *The species underlined are of American origin.*

P. engelmannii	*P. mugo*	*P. pentaphylla*	*P. rudis*
P. koraiensis	*P. muricata*	*P. pinaster*	*P. strobiformis*
P. leiophylla	*P. nigra*	*P. ponderosa*	*P. sylvestris*
P. luchuensis	*P. oocarpa*	*P. radiata*	*P. yunnanensis*
P. monticola			

From Futai and Furuno (1979) and Ishii *et al.* (1981).

Table 5. *Older trees of* Pinus *spp. killed by inoculation with* Bursaphelenchus xylophilus. *The species underlined are of American origin.*

Age of plants (y)	Species	Proportion of plants killed
13[1]	*P. densiflora*	5/5
"	*P. luchuensis*	5/5
"	*P. thunbergii*	3/3
"	*P. caribaea*	2/5
"	*P. rigida*	2/5
20[2]	*P. sylvestris*	9/15
"	*P. echinata*	1/15
"	*P. strobus*	1/9

[1] From Kiyohara and Tokushige (1971).
[2] From Linit and Tamura (1987).

mean that one out of four trees fed on by *Monochamus* carrying *B. xylophilus* would be killed. This is such a high rate that it suggests that the inoculation methods give a much higher level of mortality than might be expected. In fact, of all the species tested, I could find only two, *P. palustris* and *P. jeffreyi*, of which none of the young plants (between 1-5 years old) were killed (although even 1 or 2-month-old seedlings of the latter species could be killed (Tamura & Dropkin, 1984)).

The authors who criticised experiments with young plants suggested that more reliable results would be obtained with plants over 5 years old, but a study of experiments with plants over 5 years, and even over 20 years old, still raises some doubts. In Tables 4 and 5, a number of *Pinus* spp. are listed that have been killed by inoculation experiments using older plants as hosts and, as can be seen, they include many species other than those known to be susceptible in the field. In addition, several of

Table 6. Pinus *spp. initially reported to be killed by* Bursaphelenchus xylophilus *in the USA.*

P. banksiana	*P. elliottii*	*P. ponde*	*P. strobus*
P. cembra	*P. halepensis*	*P. ponderosa*	*P. sylvestris*
P. clausa	*P. mugo*	*P. radiata*	*P. taeda*
P. contorta	*P. nigra*	*P. resinosa*	*P. thunbergii*
P. echinata	*P. palustris*	*P. rigida*	*P. virginiana*

From Dropkin and Linit (1982).

these species are of American origin (underlined in the tables) and they are generally considered to be resistant to *B. xylophilus* (Wingfield *et al.*, 1984). In most cases where older trees were used in the experiments, the inoculation was performed by boring a hole into the trunk of the tree and pouring in a nematode suspension. This method is so unlike the natural pathway of *B. xylophilus* that it is not surprising if the results are also unlike the natural situation.

The supporting evidence that the suspected nematode species has been found in dead trees cannot be taken to indicate that the nematodes were responsible for their death. Because of the characteristic life style of the majority of *Bursaphelenchus* spp., in which they are transported to dying or very recently dead trees by insects, it is a fact that almost any dead tree, whatever its cause of death, will contain nematodes. The presence of nematodes is the result of tree death, not the cause of it. An example of how the presence of nematodes in dead trees can lead to the mistaken conclusion of pathogenicity is shown in Table 6. When the USA was surveyed after the re-discovery of *B. xylophilus* in the early 1980s, it was assumed that the nematode was responsible for the death of numerous species of *Pinus* spp. throughout the country. In fact, the predominantly saprophytic life style of *B. xylophilus* in North America was subsequently recognised and the nematode is now known not to be pathogenic to North American tree species (Wingfield *et al.*, 1984).

Discussion

The conclusions from this limited review of the literature on testing of pathogenicity of *B. xylophilus* by inoculation experiments is that the methods provide very variable results and that the overall results do not reflect the true situation in the field. Forty-one species of *Pinus* were shown in these experiments to be killed by *B. xylophilus* but, of these, only seven species are known to be killed in the field, and many species (those of American origin) are known not to be killed in the field. Only two species of *Pinus* proved not to be susceptible to the nematode. Even inoculation experiments conducted on older plants or trees in the field do not give results consistent with our knowledge of the field situation. It seems that virtually any combination of *Bursaphelenchus* sp./*Pinus* sp. tested by inoculation experiments will lead to the erroneous conclusion that the nematode is pathogenic to the host. It is not, therefore, surprising that when other *Bursaphelenchus* spp. were tested by similar methods, they also appeared to be pathogenic to the *Pinus* spp. on which they were tested.

So how could pathogenicity be demonstrated for *Bursaphelenchus* spp? The classical method in plant pathology to demonstrate that a particular organism is the causal agent of a particular disease is the use of Koch's Postulates. This is a logical process with the following steps: *i*) the pathogen must be shown to be present in every case of the disease. This requires first a characterisation of the disease to demonstrate that the symptoms are unique, recognisable and different from other diseases. *ii*) Isolation of the pathogen from the diseased host and its purification; *iii*) re-inoculation to a host of the purified pathogen, leading to the disease symptoms; and *iv*) re-isolation of the pathogen from the infected host.

Although it is now recognised that Koch's Postulates do not cover all cases of pathogen/disease associations (*e.g.*, when some non-pathogenic organisms become pathogenic under certain conditions, or if some individuals of the host species show different levels of susceptibility), they are still considered to be the essential criteria of proof in the vast majority of cases.

The first problem with regard to the use of Koch's Postulates for disease in trees supposedly caused by *Bursaphelenchus* spp. is the characterisation of the disease itself; pine wilt disease is death! Oleoresin flow stops and the leaves turn brown (and wilt in species with long pine needles). But this is not specific to pine wilt disease; these symptoms occur whenever a tree is dying from any cause. As mentioned previously, almost any dead tree, whatever its cause of death, will contain nematodes. Until the disease induced in pine trees by nematodes can be accurately characterised, this aspect of Koch's Postulates regarding association of disease and pathogen cannot be used in the case of *Bursaphelenchus* spp.

Bursaphelenchus xylophilus can be relatively easily recovered from a tree supposed to be suffering from pine wilt disease and can be cultured. In most cases, the nematodes are cultured on agar plates of fungi, and the inoculum for the experiments is obtained from this source. Very few of the published inoculation experiments have employed techniques for ensuring that the nematode inoculum was free of other organisms or other material (see Kawaza and Kaneko (1997) and Zhao *et al.* (2000), for inoculation experiments in which sterilised *B. xylophilus* did not cause mortality in young plants). Most have employed washing of the nematode suspension in sterile distilled water, but this is not equivalent to surface sterilisation and it can be assumed that, in most cases, the inoculum also included other organisms (*e.g.*, bacteria, fungi). Obviously, in order to reach the level of proof being sought by Koch's Postulates, it would be necessary to perform the inoculation with an axenic nematode suspension.

As was demonstrated above by a review of the literature, reinoculation presents a problem in attempting to prove the pathogenicity of *Bursaphelenchus* spp. For whatever reasons, the experiments so far reported do not give any confidence in the relevance of the results in relation to field conditions. There is obviously an urgent need for in-depth research into the criteria which determine the response of *Pinus* plants after inoculation with nematodes in relation to various factors, such as age of host plant, method of inoculation (including use of vector inoculation), state of sterility of the nematode suspension, life stage of the nematodes, environmental conditions, *etc.*, in order to find a means to relate such experiments to natural conditions of infection and disease expression. Therefore, without further information on disease expression and inoculation methodology, Koch's Postulates cannot be used to demonstrate the pathogenicity of *Bursaphelenchus* spp. Until this information has been obtained, the pathogenicity of these species, reported from inoculation experiments, cannot be considered to have been proved.

References

BAI, X.Y. & CHENG, H.R. (1993). The epidemical ability of pine wilt disease caused by pine wood nematode in South China. *Plant Quarantine Shanghai* 7, 333-334.

BAKKE, A., ANDERSON, R.V. & KWAMME, T. (1991). Pathogenicity of the nematodes *Bursaphelenchus xylophilus* and *B. mucronatus* to *Pinus sylvestris* seedlings: a greenhouse test. *Scandinavian Journal of Forest Research* 6, 407-412.

BEDKER, P.J. (1987). Assessing pathogenicity of the pine wood nematode. In: Wingfield, M.J. (Ed.). *Pathogenicity of the pine wood nematode*. American Phytopathological Society, St Paul, MN, USA, pp. 14-25.

BEDKER, P., WINGFIELD, M.J., BURNES, T.A., BLANCHETTE, R.A. & DROPKIN, V. (1984). Host specificity of the pine wood nematode in the North Central United States and Canada. In: Dropkin, V.H. (Ed.). *Proceedings of the United States-Japan Seminar: The resistance mechanisms of pines against pine wilt disease, 7-11 May 1984, Honolulu, Hawaii*, Columbia, MO, USA, University of Missouri, pp. 46-62.

BRAASCH, H. (1996). Pathogenitätstests mit *Bursaphelenchus mucronatus* an Kiefern- und Fichtensamlingen in Deutschland. *European Journal of Forest Pathology* 26, 205-216.

BRAASCH, H. (1997). Wirts- und Pathogenitätsuntersuchungen mit dem Kiefernholznematoden (*Bursaphelenchus xylophilus*) aus Nordamerika unter mitteleuropaischen Klimabedingungen. *Nachrichtenblatt des Deutschen Pflanzenschutzdienstes* 49, 209-214.

BRAASCH, H., CAZROPPO, S., AMBROGIONI, L., MICHALOPOULOS, H., SKARMOUTSOS, G. & TOMICZEK, C. (1998). Pathogenicity of various *Bursaphelenchus* species to pines and implications to European forests. In: Futai, K., Togashi, K. & Ikeda, T. (Eds). *Sustainability of pine forests in relation to pine wilt and decline. Proceedings of the symposium, Tokyo, Japan, 26-30 October 1998*. Kyoto, Japan, Shokado Shoten, pp. 14-22.

BURNES, T.A., WINGFIELD, M.J., BAKER, F.A., KNOWLES, K. & BEAUBIEN, Y. (1985). The pine wood nematode in jack pine infected with dwarf mistletoe. *Forest Science* 31, 866-870.

CHANG, R.J. & CHANG, R.J. (1999). Effects of inoculation method and inoculum concentration on the occurrence of pine wilt disease. *Taiwan Journal of Forest Science* 14, 409-417.

CHENG, H.R., LIN, M.S. & Q, R.J. (1986). [A study on the morphological diagnosis and the pathogenicity of *Bursaphelenchus mucronatus*.] *Journal of Nanjing Agricultural University* 2, 55-61.

DROPKIN, V.H. & LINIT, M.J. (1982). Pine wilt – a disease you should know. *Journal of Arboriculture* 8, 1-6.

DROPKIN, V.H., FOUDIN, A., KONDO, E., LINIT, M.J., SMITH, M. & ROBBINS, K. (1981). Pinewood nematode: a threat to US forests? *Plant Disease* 65, 1022-1027.

DWINELL, D.L. (1985). Relative susceptibilities of five pine species to three populations of the pinewood nematode. *Plant Disease* 69, 440-442.

DWINELL, D.L. (1997). The pinewood nematode: regulation and mitigation. *Annual Review of Phytopathology* 35, 153-166.

EVANS, H.F., MCNAMARA D.G., BRAASCH, H., CHADOEUF, J. & MAGNUSSON, C. (1996). Pest risk analysis (PRA) for the territories of the European Union (as PRA area) on *Bursaphelenchus xylophilus* and its vectors in the genus *Monochamus*. *EPPO Bulletin* 26, 199-249.

FUTAI, K. & FURUNO, T. (1979). [The variety of resistances among pine species to pine wood nematode, *Bursaphelenchus lignicolus*.] *Bulletin of Kyoto University Forests* 51, 23-36.

ISHII, K., KORINOBU, S., OHBA, K. & FURUKOSHI, T. (1981). [Resistance of interspecific hybrids among the pine species belonging to Sylvestris subsect. to pine wood nematode, *Bursaphelenchus lignicolus*.] *Transactions of the 92nd Meeting of the Japanese Forestry Society*, pp. 291-292.

KAWAZU, K. & KANEKO, N. (1997). Asepsis of the pine wood nematode isolate OKD-3 causes it to lose its pathogenicity. *Japanese Journal of Nematology* 27, 76-80.

KANEKO, S. & ZINNO, Y. (1986). Development under different light sources of pine-wilt disease caused by *Bursaphelenchus xylophilus* on the seedlings of Japanese red pine. *Journal of the Japanese Forestry Society* 68, 208-209.

KISHI, Y. (1999). Influence of tree age on wilt and mortality of pines after inoculation with *Bursaphelenchus xylophilus*. *Journal of the Japanese Forestry Society* 81, 330-333.

KIYOHARA, T. & TOKUSHIGE, Y. (1971). [Inoculation experiments of a nematode, *Bursaphelenchus* sp., onto pine trees.] *Journal of the Japanese Forestry Society* 53, 210-218.

LEE, M.J. (1986). Resistance of pine species in Taiwan to pine-wood nematodes. *Quarterly Journal of Chinese Forestry* 19, 27-33.

LINIT, M.J. & TAMURA, H. (1987). Relative susceptibility of four pine species to infection by pinewood nematode. *Journal of Nematology* 19, 44-50.

MAMIYA, Y. & KIYOHARA, T. (1972). Description of *Bursaphelenchus lignicolus* n. sp. (Nematoda: Aphelenchoididae) from pine wood and histopathology of nematode-infested trees. *Nematologica* 18, 120-124.

MAMIYA, Y. (1983). Pathology of the pine wilt disease caused by *Bursaphelenchus xylophilus*. *Annual Review of Phytopathology* 21, 201-220.

MYERS, R.F. (1984). Comparative histology and pathology in conifers infected with pine wood nematode, *Bursaphelenchus xylophilus*. In: Dropkin, V.H. (Ed.). *Proceedings of the United States-Japan Seminar: The resistance mechanisms of pines against pine wilt disease, 7-11 May 1984, Honolulu, Hawaii*, Columbia, MO, USA, University of Missouri, pp. 91-97.

OHBA, K., FUKOKOSHI, T., KURINOBU, S. & ISHII, K. (1984). Susceptibility of subtropical pine species and provenance to the pine wood nematode. *Journal of the Japanese Forestry Society* 66, 465-468.

RIGA, E., SUTHERLAND, J.R. & WEBSTER, J.M. (1991). Pathogenicity of pinewood nematode isolates and hybrids to Scots pine seedlings. *Nematologica* 37, 285-292.

SHAUER-BLUME, M. (1990). Preliminary investigations on pathogenicity of European *Bursaphelenchus* species in comparison to *Bursaphelenchus xylophilus* from Japan. *Revue de Nématologie* 13, 191-195.

SKARMOUTSOS, G. & SKARMOUTSOS-MICHALOPOULOS, H. (2000). Pathogenicity of *Bursaphelenchus sexdentati, Bursaphelenchus leoni* and *Bursaphelenchus hellenicus* on European pine seedlings. *Forest Pathology* 30, 149-156.

SUTHERLAND, J.R., RING, F.M. & SEED, J.E. (1991). Canadian conifers as hosts of the pinewood nematode (*Bursaphelenchus xylophilus*): results of seedling inoculations. *Scandinavian Journal of Forest Research* 6, 209-216.

TAMURA, H. & DROPKIN, V. (1984). Resistance of pine trees to pine wilt caused by the nematode, *Bursaphelenchus xylophilus. Journal of the Japanese Forestry Society* 66, 306-312.

WINGFIELD, M.J. & BLANCHETTE, R.A. (1984). Pathogenicity and insect associates of the pine wood nematode in the North Central States In: Dropkin, V.H. (Ed.). *Proceedings of the United States-Japan Seminar: The resistance mechanisms of pines against pine wilt disease, 7-11 May 1984, Honolulu, Hawaii*, Columbia, MO, USA, University of Missouri, pp. 32-45.

WINGFIELD, M.J. (1987). A comparison of the mycophagous and phytophagous phases of the pine wood nematode. In: Wingfield, M.J. (Ed.). *Pathogenicity of the pine wood nematode.* American Phytopathological Society, St Paul, MN, USA, pp. 81-90.

WINGFIELD, M.J., BLANCHETTE, R.A. & KONDO, E. (1983). Comparison of the pine wood nematode, *Bursaphelenchus xylophilus* from pine and balsam fir. *European Journal of Forest Pathology* 13, 360-372.

WINGFIELD, M.J., BLANCHETTE, R.A. & NICHOLLS, T.H. (1984). Is the pine wood nematode an important pathogen in the United States? *Journal of Forestry* 82, 232-235.

YANG, B.J., WANG, Q.L., ZOU, W.D. & LI, Y.Z. (1988). Study on the pathogenicity of *Bursaphelenchus mucronatus* to pines. *Forest Science and Technology* 1, 21-23.

YANG, B.J. *et al.* (1987). [The resistance of pines species to pine wood nematode, *Bursaphelenchus xylophilus*.] *Acta Phytopathologica Sinica* 17, 211-214.

ZHAO, B.G., GAO, R., JU, Y.W., GUO, D.S., GUO, J., ZHAO, B.G., GAO, R., JU, Y.W., GUO, D.S. & GUO, J. (2000). Effects of antibiotics on pine wilt disease. *Journal of Nanjing Forestry University* 24, 75-77.

Nematology Monographs & Perspectives, 2003, Vol. 1, 199-214

The effects of simulated acid rain on the development of pine wilt disease

Ei-ichiro ASAI and Kazuyoshi FUTAI

Laboratory of Environmental Mycoscience, Graduate School of Agriculture, Kyoto University, Sakyo-ku, Kyoto 606-8502, Japan

Summary – Six-month-old Japanese black pine seedlings were exposed to simulated acid rain (SAR) at pH 3 and 2, three times a week. After this treatment for 2 months, the seedlings were inoculated with virulent (S10) or avirulent (C14-5) isolates of pine wood nematode at three inoculum densities (50, 160 or 500 nematodes per seedling). In the seedlings inoculated with 500 virulent nematodes, disease development was accelerated by the pretreatment with SAR at pH 2 or 3. In the seedlings inoculated with 50 virulent nematodes, however, the appearance of dead seedlings was retarded by the pretreatment with pH 3 SAR. SAR at pH 2 increased seedling mortality irrespective of the number of virulent nematodes inoculated. When inoculated with 500 avirulent nematodes, SAR at pH 2 or 3 slightly increased seedling mortality. When inoculated with 50 or 160 avirulent nematodes, almost all seedlings survived throughout the experimental period except for the seedlings pretreated with SAR at pH 2. The relationship between the cumulative mortality of the seedlings and the load on the host given by nematodes, the product of the logarithm of the number of nematodes inoculated (*Pi*) and the time after inoculation (t), was analysed by linear regression. The results suggest that SAR at pH 3 not only raised the rate of increase in seedling mortality, but also increased resistance of pines to the virulent nematodes. SAR at pH 3 is suggested to sustain little damage to the resistance of pines to the avirulent nematodes. According to these results, we conclude that acid rain at pH 2 decreases the resistance of pines to pine wood nematodes, and acts as a promoting factor of pine wilt disease even if pines were not directly killed by the impacts of acid rain. Acid rain at the current ambient level in Japan, however, may not increase the mortality of pine trees infected by pine wood nematodes, but rather retard the appearance of dead trees.

Almost 30 years have passed since pine wood nematode, *Bursaphelenchus xylophilus* (Steiner & Buhrer) Nickle was discovered to be the pathogen of pine wilt disease (Kiyohara & Tokushige, 1971). Since then, enormous efforts have been made to control this epidemic forest disease in Japan; most of the efforts, however, seem to be fruitless. Because of unsuccessful results in control procedures, some scientists have proposed air pollutants as the alternative causal agents of the widespread devastation of pine forests (Taniyama, 1989). Among them, some extremists have put too much stress on the effect of air pollution, and underestimated (or even neglected) the effect of pine wilt disease, which has prevented constructive discussion based on scientific data. For this reason, only a limited number of studies have been conducted to reveal the role of air pollutants in developing the disease so far (Tanaka, 1975; Bolla & Fitzsimmons, 1988; Futai & Harashima, 1990; Taketsune, 1992). Among the air pollutants today, acid rain is especially common to most developed countries and one of the causes of forest decline in European and North American countries (Morrison, 1984). We therefore began to conduct experimental studies to know whether or not acid rain affects the progress of this epidemic disease and, if so, how.

Effects of acid rain on the development of pine wilt disease are rather complex. For example, repeated exposure to simulated acid rain (SAR) at pH 3 retarded the appearance of symptoms after nematode inoculation, although the seedlings pretreated with SAR ceased resin exudation from the stem earlier than those pretreated with tap water (Asai & Futai, 2001a). These results imply that exposure to acid rain not only increases the damage to pines caused by the nematodes (pine wilt disease), but delays development of the disease. The first aim of our present study is to examine whether or not acid rain has such dual effects on the development of pine wilt disease. In our previous experiment, the development of disease symptoms in 4-month-old Japanese black pine seedlings inoculated with 500 pine wood nematodes was accelerated by pretreatment with SAR at either pH 4, 3, or 2 (Asai & Futai, 2001b). This result suggests that the inoculum density of 500 nematodes per seedling imposes too heavy a burden on juvenile seedlings to show dual effects of acid rain. In the first experiment (Expt 1), therefore, we examined the effects of pretreatment with SAR at pH 2 and 3 on the development of pine wilt disease in pine seedlings inoculated with pine wood nematodes at a lower density (50, 160 and 500 nematodes per seedling), and thereby tried to elucidate the influence of dose-response relationships between pines and pine wood nematodes on the result of acid rain treatment.

Most forest pathologists in Japan have considered that the serious epidemic death of pine forests throughout Japan is mainly due to a high susceptibility of Japanese native pine species to pine wilt disease, and have disregarded the role of air pollution, including acid rain, as a promoting factor of the disease. However, it is a well-known fact that populations of pine wood nematode in the field consist of many isolates with variable virulence to pines (Kiyohara & Bolla, 1990). It is also well known that pines under severe stress are killed by infection of avirulent isolates of pine wood nematode (Ikeda, 1996). Thus, it is possible that latent damage to the pines by air pollution might destroy or weaken their resistance to avirulent pine wood nematode. Tanaka (1975) reported that fumigation with SO_2 at 0.2 ppm for 10 weeks increased the mortality of Japanese black pine seedlings inoculated with *B. mucronatus*, the species related to pine wood nematode but which has little pathogenicity to pines. In the second experiment (Expt 2), therefore, we inoculated 1-year-old Japanese black pine seedlings pretreated with SAR at pH 3 and 2 with an avirulent isolate of pine wood nematode, with 50, 160 or 500 nematodes per seedling.

Materials and methods

Seedlings

Seeds of Japanese black pine, *Pinus thunbergii* Parl., sterilised with sodium hypochlorite were sown on sterilised, wet paper towel (Expt 1) or on vermiculite (Expt 2). About 2-5 weeks after sowing, germinated seeds were planted in plastic cylindrical vials (4 cm diam., 11.2 cm in height, with a hole of 3.5 mm diam. at the bottom) filled with 100 cm^3 of autoclaved vermiculite, and were grown in the greenhouse until inoculation with pine wood nematodes. About 1600 juvenile seedlings in total served for the model plants in this study. The dates for sowing seeds, planting germinated seeds, and commencement and ending of SAR pretreatment in each experiment, are described in Table 1. The seedlings were grown under a natural photoperiod and poorly controlled air temperature in the greenhouse (Table 1).

Exposure to SAR

An acidic solution was prepared by mixing 0.5 M sulphuric acid and nitric acid at an S : N ratio of 3 : 1, and the solution was diluted with distilled water (DIW) or tap water (TW), then adjusted to pH 3 or 2. In

Table 1. *The details of Experiments 1 and 2. Start: start of simulated acid rain (SAR) pretreatment; End: end of SAR pretreatment. Environmental conditions (maximum and minimum temperature and daylength) in the greenhouse during the SAR pretreatment are also described.*

	Experiment 1	Experiment 2
Sowing	13 Feb. 1999	29 Apr. 1999-30 May 1999
Planting	26 Feb. 1999	15 May 1999-25 Jun. 1999
Start	19 Aug. 1999	13 Dec. 1999
End	24 Oct. 2000	16 Feb. 2000
Max. temp.	32 ± 2°C	22 ± 2°C
Min. temp.	18 ± 2°C	9 ± 2°C
Daylength	11:01-13:25	9:45-10:55

Table 2. *Weekly schedule of the simulated acid rain (SAR) treatment. DIW: distilled water, TW: tap water. The seedlings were treated with SAR applied as a spray to the top alone (T) or by irrigation of both the top and roots (TR) at pH 2 and 3.*

Treatment	Mon. spray	Wed. spray	Fri. spray	Sun. irrigation
W (control)	DIW	DIW	DIW	TW
T (pH 2)	SAR	SAR	SAR	TW
TR (pH 2 and 3)	DIW	SAR	SAR	SAR

this study, based on the results of our preliminary experiment (Asai & Futai, 2001b), the seedlings were divided into the following four groups and were treated with SAR or water according to the schedule shown in Table 2. In the pH 2 T group, only the tops of the seedlings were sprayed with pH 2 SAR three times a week. In the pH 2 TR and pH 3 TR groups, the tops of the seedlings were sprayed with SAR (pH 2 and 3, respectively) three times a week, and the roots were also exposed to SAR (pH 2 and 3, respectively) once a week. DIW was used to prepare SAR for foliar spray, and TW to prepare SAR for application to roots. In all groups SAR or DIW was sprayed until the needles were wetted to dripping point (about 1.28 ml per seedling), and about 20.2 ml of SAR or TW was applied to the roots of each seedling.

During the SAR treatment period, we sometimes watered the seedlings with TW to keep the soil moist, and transposed once a week to give uniform treatment for all of the seedlings. In total, we prepared 640 seedlings in Expt 1 and 844 seedlings in Expt 2; thus, each of the experimental groups in Expts 1 and 2 contained 160 and 211 seedlings, respectively.

Nematode inoculation

Inoculation with pine wood nematode was carried out the day after ending the pretreatment with SAR. A virulent and an avirulent isolate of pine wood nematode, S10 and C14-5, were used for the inoculation tests in Expts 1 and 2, respectively. In both experiments, 135 seedlings arbitrarily selected from each group were divided into three subgroups of 45 seedlings each, and each subgroup was inoculated with 50, 160, or 500 pine wood nematodes. In Expt 2, in addition to the inoculation with avirulent nematodes, 15 seedlings from each experimental group were inoculated with 500 virulent pine wood nematodes as a positive control. The remaining seedlings in each group ($n = 25$ in Expt 1 and $n = 15$ in Expt 2) were inoculated with tap water as a negative control. For the inoculation, we made a small slit on the stem with a knife, inserted a small triangular piece of filter paper into the slit, and pipetted a 20 μl water suspension of nematodes or the same amount of tap water onto each filter paper.

After the inoculation, seedlings were moved to two growth chambers ($120 \times 60 \times 55$ cm), and covered with transparent vinyl sheet for 24 h to prevent drying. In the chambers, all of the planted vials were placed in plastic vats ($36 \times 47 \times 3$ cm) filled with water to keep the soil moist. Each growth chamber was installed with four daylight balance fluorescent lamps, and the plants were exposed to light at about 10 W/m^2 from 07:00 to 23:00 h (16 h photoperiod). The temperature was kept at 23.8-25.8°C during the light period and 19.1-20°C during the dark period. The relative humidity was 65-70%.

Observation of disease development

Visual symptoms caused by pine wilt disease can be easily distinguished from the necrosis caused by SAR which mainly appears as spot- or stripe-shaped wounds usually from the tip of needles, and sometimes extends all over the needles except for the needle base, while the necrosis due to pine wilt disease exhibits symptoms from the basal part of the needles (Fig. 1). When the symptoms such as necrosis, discoloration, or

Fig. 1. *Needle necrosis in Japanese black pine seedlings caused by the exposure to simulated acid rain at pH 2 (upper) and needle discoloration in the seedlings caused by pine wilt disease (lower). The right seedling in the lower photograph is healthy.*

drooping of the needles extended to the tip of the stem, the seedling was regarded as dead. The mortality of the seedlings was recorded every 3rd day from the 12th day in Expt 1, and from the 14th day in Expt 2.

Data analysis

The effect of SAR on the mortality of pine seedlings after nematode inoculation was analysed as follows. Assuming that the population of pine wood nematodes increases exponentially with the time elapsed, cumulative values of load given to the host by the nematodes (L) can be represented as the product of log (Pi) and t, and cumulative mortality of the seedlings (M) will positively correlate with L because M increases

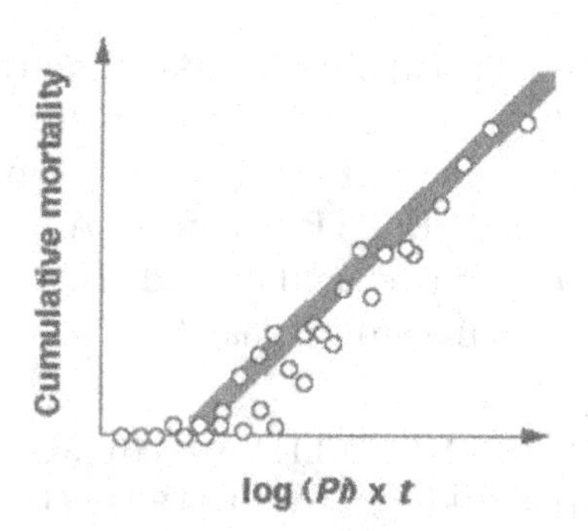

Fig. 2. *Method for analysing the effects of simulated acid rain on the relationship between cumulative stress (log(*Pi*) × *t*) after infection by pine wood nematode (PWN) and seedling mortality.* Pi: *initial population density of PWN,* t: *time after inoculation. The slope (*S*) and the x-intercept (*To*) of the regression line reflects the rate of increase in the seedling mortality, and the critical load necessary to kill the seedlings, respectively.*

with the time (Fig. 2). L and M were fitted to the linear regression line as follows:

$$M = -a + SL$$

where a and S are constants, the gradient of the regression line (S) shows a rate of increase in mortality. On the other hand, the x-intercept of the regression line (*To*) shows the critical value of ($\log(Pi) \times t$) needed to kill the seedling, and is thought to reflect the resistance of the seedling to the infection of pine wood nematodes. From the above equation, *To* equals to the quotient of a divided by S. The values of S and *To* obtained from each treatment were compared among the groups or subgroups.

Results

Effect of SAR on Japanese black pine seedlings

Necrosis appeared on the needles of Japanese black pine seedlings only when exposed to pH 2 SAR, and no symptoms appeared on the seedlings exposed to pH 3 SAR. Most of the seedlings not inoculated with nematodes were alive at the end of the experiment irrespective of the exposure to SAR in Expt 1. The mortality of the seedlings without nematode inoculation at the end of the experimental period in Expt 1 was 3.1, 0.6, 0.6, and 3.8% in W, pH 3 TR, pH 2 T, and pH 2 TR treatment, respectively. In Expt 2, the exposure to SAR slightly (but significantly) increased the mortality of the seedlings before inoculation. The mortality at the end of SAR treatment in Expt 2 was 0.5, 6.6, 10.4 and 2.4% in W, pH 3 TR, pH 2 T, and pH 2 TR treatment. All of the seedlings that received no nematodes survived until the end of the experimental period after they were moved to the growth chambers.

EFFECT OF SAR ON THE MORTALITY OF THE SEEDLINGS AFTER INOCULATION WITH A VIRULENT ISOLATE OF PINE WOOD NEMATODES

Dead seedlings appeared on the 12th day after inoculation, and increased thereafter with time (Fig. 3). The larger the *Pi*, the more promptly mortality increased. In the control group (W), mortality of the seedlings after inoculation with 50, 160 and 500 nematodes was 46.7, 57.1 and 71.4%, respectively, at the end of the experiment. Mortality was always more in the pH 2 TR group than in the W group irrespective of *Pi*. In the pH 3 TR group, however, mortality after inoculation with 50 or 160 nematodes was less than that in the W group during the first 20-25 days after inoculation, and the dead seedlings appeared later than in the W group (Fig. 3). When inoculated with 500 nematodes, mortality

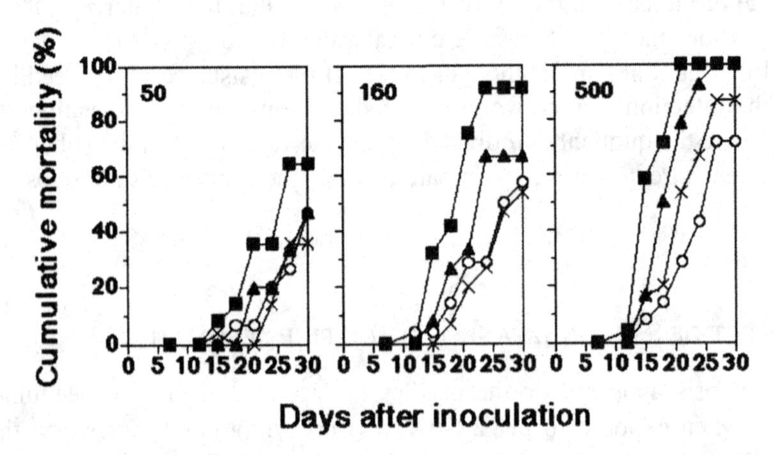

Fig. 3. *Changes in cumulative mortality of Japanese black pine seedlings, pretreated by applying simulated acid rain at pH 3 and 2 to leaves and to leaves and roots, after inoculation with virulent pine wood nematode (PWN) at 50, 160 or 500 per plant, ○: SAR control (water); ×: SAR at pH 3 to leaves and roots; ▲: SAR at pH 2 to leaves; ■: SAR at pH 2 to leaves and roots.*

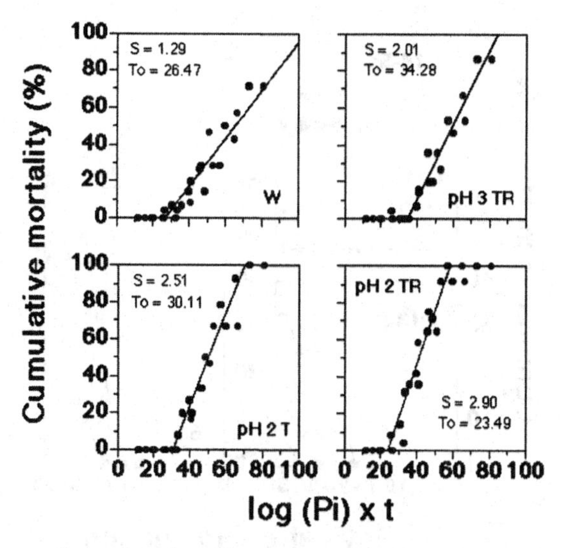

Fig. 4. *Effects of simulated acid rain on mortality of Japanese black pine seedlings after inoculation with a virulent pine wood nematode isolate. The slope (*S*) and the x-intercept (*To*) of the regression line reflects the rate of increase in the seedling mortality, and the critical load necessary to kill the seedlings, respectively.*

of the seedlings was greater in the pH 3 TR group than in the W group 15 days after inoculation.

The cumulative mortality of the seedlings in each subgroup plotted against $\log(Pi) \times t$, fitted well to a linear regression line (Fig. 4). The equation of linear regression in each treatment is as follows: $Y = 1.29X - 34.22$ (W), $Y = 2.01X - 69.03$ (pH 3 TR), $Y = 2.51X - 75.62$ (pH 2 T), and $Y = 2.90X - 68.05$ (pH 2 TR). The value S, the gradient of the regression line, increased as the SAR level applied to the seedlings increased (Fig. 4). On the other hand, *To*, the x-intercept, in the pH 3 TR and pH 2 T groups was larger than that in the control (W group), and *To* in the pH 2 TR group was the smallest among the four groups (Fig. 4).

Effect of SAR on the mortality of the seedlings after inoculation with an avirulent isolate of pine wood nematodes

Almost all seedlings survived when inoculated with 50 or 160 avirulent pine wood nematodes except for a total of two seedlings in pH 2 T treatment (Fig. 5). When inoculated with 500 virulent or avirulent pine

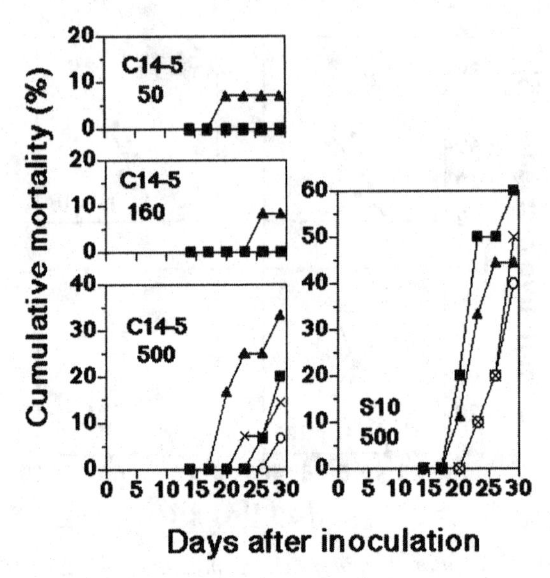

Fig. 5. *Cumulative mortality of Japanese black pine seedlings, pretreated by applying simulated acid rain at pH 3 and 2 to leaves and to leaves and roots, after inoculation with an avirulent pine wood nematode (PWN) isolate C14-5 at 50, 160 or 500 per plant or with a virulent PWN isolate S10 at 500 per plant.* ○*: SAR control (water);* ×*: SAR at pH 3 to leaves and roots;* ▲*: SAR at pH 2 to leaves;* ■*: SAR at pH 2 to leaves and roots.*

wood nematodes, the mortality of the seedlings was slightly increased by the exposure to SAR. Mortality of the seedlings apparently differed between virulent and avirulent isolates irrespective of four simulated rain treatments. Final mortality in the seedlings in W treatment was 6.6% when inoculated with 500 avirulent nematodes; on the other hand, it amounted to 40% in total when inoculated with 500 virulent nematodes (Fig. 5).

Figure 6 shows the cumulative mortality of the seedlings in each subgroup plotted against $\log(Pi) \times t$. The equation of linear regression in each treatment is as follows: $Y = 0.19X - 10.78$ (W), $Y = 0.48X - 25.15$ (pH 3 TR), $Y = 0.67X - 22.96$ (pH 2 T), and $Y = 0.67X - 37.87$ (pH 2 TR). When the values of S and To among four experimental groups were compared with each other as Expt 1, value S was also the smallest in W treatment, and also increased with SAR level applied to the seedlings. On the other hand, To of pH 2 T treatment was the smallest

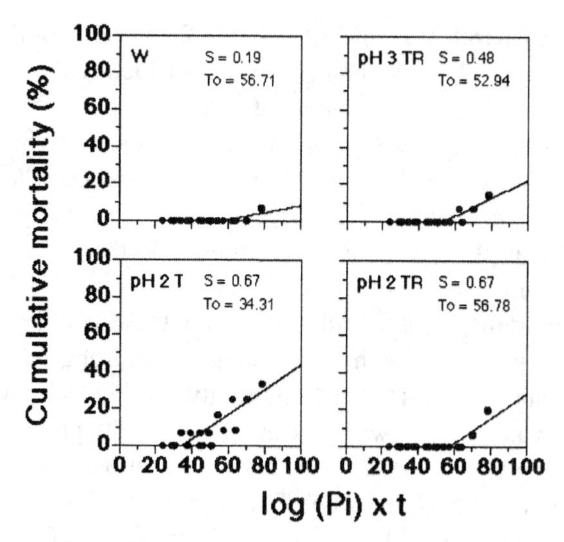

Fig. 6. *Effects of simulated acid rain on mortality of Japanese black pine seedlings after inoculation with an avirulent pine wood nematode isolate. The slope (S) and the x-intercept (*To*) of the regression line reflects the rate of increase in the seedling mortality, and the critical load necessary to kill the seedlings, respectively.*

among the four treatments, and the values of the other three treatments (W, pH 3 TR, and pH 2 TR) were almost equal (Fig. 6).

Discussion

In Expt 1, most of the seedlings receiving no nematodes survived until the end of the experiment irrespective of the exposure to SAR; on the other hand, the exposure to SAR in Expt 2 killed a maximum 10.4% of seedlings without nematode inoculation. This is probably because environmental conditions in the greenhouse were not constantly controlled. The increase in seedling mortality by SAR in Expt 2 might depend on the seasonal factor(s) specific to the winter, such as low air temperature and short day length. Many investigators have demonstrated that cold tolerance of conifers is reduced by exposure to air pollutants including acid deposition (Sheppard *et al.*, 1993; Esch & Mengel, 1998). In Expt 2, the necrosis in the needles occurred more often and more heavily on the seedlings pretreated with pH 2 T than those with pH 2 TR, which was inconsistent with the result of Expt 1 (data not shown). When inoculated with avirulent nematodes in Expt 2, mortality of the seedlings pretreated with pH 2 T was more than that of the seedlings with pH 2 TR. The reason for the improvement of damage in pH 2 TR treatment is obscure. However, Sheppard *et al.* (1993) reported that simulated acid mist at pH 2.7 reduced frost hardiness of red spruce (*Picea rubens*) seedlings, but the reduction of frost hardiness was ameliorated when sulphuric acid at pH 2.7 was exposed directly to the soil in addition to simulated acid mist.

Seedling mortality after inoculation with 500 virulent nematodes was more in all groups treated with SAR than in the control group throughout the experimental period. However, mortality after inoculation with 160 or 50 virulent nematodes was less in the pH 3 TR group than in the W group during the first 20-25 days after inoculation. Linear regression analysis (Fig. 4) showed that value S (rate of increase in mortality) was higher in either the pH 3 TR or pH 2 T group than in the W (control) group. However, the value To (the minimum value of $(\log(Pi) \times t)$ necessary to kill a seedling) was also larger in both pH 3 TR and pH 2 T groups than in the control group. These results suggest that these two pretreatments with SAR not only raised the rate of increase in mortality, but also increased the resistance of seedlings to the nematodes.

In Expt 1, To was smaller and the S was more in the pH 2 TR group than in the control group. The cumulative mortality was also more in the pH 2 TR group than that in the control group throughout the experimental period. Pretreatment with SAR at pH 2 T decreased the value To in Expt 2, which resulted in the death of seedlings in pH 2 T treatment at lower inoculum density level (50 and 160 nematodes). These results imply that exposure to pH 2 SAR decreased the resistance of seedlings to the nematodes.

Pretreatment with SAR raised the rate of increase in seedling mortality (S) after inoculation with avirulent nematodes as in the case of virulent nematodes. However, value S of the seedlings in Expt 2 was far smaller than that in Expt 1. This is mainly due to the lower population growth of avirulent pine wood nematodes in Japanese black pine seedlings than that of virulent nematodes (data not shown). Many investigators have pointed out that the virulence of pine wood nematode is parallel to the population growth rate in seedlings (Kiyohara, 1989). Our present study also indicated that acid rain at pH 3 and 2 increased the population growth of pine wood nematodes in the seedlings (data not shown).

When inoculated with 160 avirulent nematodes in Expt 2, nematode population growth in seedlings pretreated with SAR at pH 2 or 3 was significantly larger than in that in the control (W) seedlings (data not shown). Only one seedling in pH 2 T treatment, however, was killed by nematode inoculation. This suggests that pines can survive even if acid rain increases the population growth of avirulent nematodes to a certain extent. In other words, avirulent nematodes do not appear to be a serious threat to the pines. Such a relationship between avirulent nematodes and Japanese black pine is called 'weakly pathogenic' or 'endemic', whereas the relationship between virulent nematode and Japanese black pine is 'pathogenic' or 'epidemic'.

Evidence strongly suggests that pine wood nematode originated in North America (Dropkin, 1984; Iwahori *et al.*, 1998). Pine wood nematodes in North America cause tree death only occasionally (Dropkin, 1984) because pine species native to that place, such as loblolly pine (*P. taeda*) and slash pine (*P. elliottii*), have a strong resistance to the nematodes (Futai & Furuno, 1979). Generally, *Bursaphelenchus* species are saprophagous organisms or facultative parasites that can kill trees in decline only. This aspect of the nature of pine wood nematode might be reflected in the activation of population growth in the seedlings pretreated with SAR. Contrary to facultative parasites, acid rain might have a negative effect on the amenity of plants for obligate parasites. As for obligate-parasitic nematodes, several investigators reported that reproduction of *Meloidogyne incognita*, *M. hapla* and *Heterodera glycines* was suppressed by acid rain at pH 2.3-3.2 (Shriner, 1978; Shafer *et al.*, 1992; Khan & Khan, 1994). Pine species native to East Asia generally have little resistance to infection by pine wood nematodes. For example, six to ten Japanese red pine (*P. densiflora*) trees were killed by inoculation with only 30 nematodes (Kiyohara *et al.*, 1973). Value *To* to virulent nematodes in Expt 1 is far smaller than that to avirulent nematodes in Expt 2, which shows that pine wood nematodes have acquired the ability to infect healthy pine seedlings. The role of air pollutants, including acid rain, on the spread of epidemic death of pine trees in Japan has been discussed since the 1970s. Inoculation of pine wood nematode at various densities is one of the methods available to examine the effects of environmental stresses, such as acid rain, on the incidence of this disease. The present study suggested that exposure to pH 3 SAR not only raised the rate of mortality, but also increased critical load necessary to kill the seedlings; and that exposure to pH 2 SAR raised the rate of mortality and hastened the appearance of dead seedlings. Our previous experiment (Asai & Futai, 2001) suggested that acid rain at pH 3 improved resistance of Japanese black pine seedlings to pine wood nematode, which supports the results of present study.

Therefore, we conclude that acid rain at the current ambient level in Japan (the maximum acidity of rain observed at the Kamigamo Experimental Station of Kyoto University Forests was pH 3.4 (Kaneko *et al.*, 1993)) may not increase mortality of pine trees infected by pine wood nematodes, but rather retard the appearance of dead trees. In the previous study, however, we indicated that even acid rain at pH 4 has the potential to promote population growth of pine wood nematode in Japanese black pine (Asai & Futai, 2001b). The present study does not eliminate the possibility that acid rain promotes the incidence of pine wilt disease in collaboration with other environmental factor(s) such as high air temperature and drought, even though acid rain itself does not kill pines and, alone, does not act as a promoting factor of pine wilt disease. We think it is quite possible that in the near future pine forests in developing countries could be ruined by the interaction of pine wood nematode invasion with damage from heavy acid rain, because of insufficient countermeasures against air pollution, including acid rain, by almost all these countries. Pine wilt disease, as well as acid rain, is a worldwide environmental problem caused by the globalisation of human activity.

References

ASAI, E. & FUTAI, K. (2001a). Retardation of pine wilt disease symptom development in Japanese black pine seedlings exposed to simulated acid rain and inoculated with *Bursaphelenchus xylophilus*. *Journal of Forest Research* 6, 297-302.

ASAI, E. & FUTAI, K. (2001b). The effects of long-term exposure to simulated acid rain on the development of pine wilt disease caused by *Bursaphelenchus xylophilus*. *Forest Pathology* 31, 241-253.

BOLLA, R.I. & FITZSIMMONS, K. (1988). Effect of simulated acid rain in *Bursaphelenchus xylophilus* infection of pine seedlings. *Journal of Nematology* 20, 590-598.

DROPKIN, V.H. (1984). Pine wilt in the U.S. In: Dropkin, V.H. (Ed.). *Proceedings of the United States-Japan Seminar; The resistance mechanisms of pines against pine wilt disease, 7-11 May 1984, Honolulu, Hawaii.* Columbia, MO, USA, University of Missouri, pp. 3-5.

ESCH, A. & MENGEL, K. (1998). Combined effects of acid mist and frost drought on the water status of young spruce trees (*Picea abies*). *Environmental Experimental Botany* 39, 57-65.

FUTAI, K. & FURUNO, T. (1979). [The variety of resistances among pine species to pine wood nematode, *Bursaphelenchus lignicolus*.] *Bulletin of the Kyoto University Forests* 51, 23-36.

FUTAI, K. & HARASHIMA, S. (1990). Effect of simulated acid mist on pine wilt disease. *Journal of Japanese Forestry Society* 72, 520-523.

IKEDA, T. (1996). Responses of water-stressed *Pinus thunbergii* to inoculation with avirulent pine wood nematode (*Bursaphelenchus xylophilus*): water relations and xylem histology. *Journal of Forest Research* 1, 223-226.

IWAHORI, H., TSUDA, K., KANZAKI, N., IZUI, K. & FUTAI, K. (1998). PCR-RFLP and sequencing analysis of ribosomal DNA of *Bursaphelenchus* nematodes related to pine wilt disease. *Fundamental and Applied Nematology* 21, 655-666.

KANEKO, T., NAKAI, I. & ANDO, M. (1993). [Measurement of acid rain using commercially sold acid rain collector – changes of pH at the beginning of a rain.] *Transactions of Kansai Branch of the Japanese Forestry Society* 2, 57-58.

KHAN, M.R. & KHAN, M.W. (1994). Effects of simulated acid rain and root-knot nematode on tomato. *Plant Pathology* 43, 41-49.

KIYOHARA, T. (1989). [Etiological study of pine wilt disease.] *Bulletin of Forest and Forestry Products Research Institute* 353, 127-176.

KIYOHARA, T. & BOLLA, R.I. (1990). Pathogenic variability among populations of the pinewood nematode, *Bursaphelenchus xylophilus*. *Forest Science* 36, 1061-1076.

KIYOHARA, T. & TOKUSHIGE, Y. (1971). Inoculation experiment of a nematode, *Bursaphelenchus* sp., onto pine trees. *Journal of Japanese Forestry Society* 53, 210-218.

KIYOHARA, T., DOHZONO, Y., HASHIMOTO, H. & ONO, K. (1973). [Correlation between number of inoculated pinewood nematodes and disease occurrence in pine wilt disease.] *Transactions of the Meeting of Kyushu Branch of the Japanese Forestry Society* 26, 191-192.

MORRISON, I.K. (1984). Acid rain – a review of literature on acid deposition effects in forest ecosystems. *Forestry Abstracts* 45, 483-506.

SHAFER, S.R., KOENNING, S.R. & BARKER, K.R. (1992). Interactions of simulated acidic rain with root-knot or cyst nematodes on soybean. *Phytopathology* 82, 962-970.

SHEPPARD, L.J., CAPE, J.N. & LEITH, I.D. (1993). Influence of acidic mist on frost hardiness and nutrient concentrations in red spruce seedlings: 1. Exposure of the foliage and the rooting environment. *New Phytologist* 124, 595-605.

SHRINER, D.S. (1978). Effects of simulated acid rain on host-parasite interactions in plant diseases. *Phytopathology* 68, 213-218.

TAKETSUNE, A. (1992). [Effects of acid rain and acid fog on pine wilt disease. (I) Detection of pine wood nematode (*Bursaphelenchus xylophilus* Steiner & Brer) from dead trees in the foggy district and tests with simulated acid water sprayed to seedlings inoculated with this nematode.] *Bulletin of the Hiroshima Prefectural Forest Experiment Station* 26, 23-30.

TANAKA, K. (1975). [The effect of sulphur dioxide on the infection of pine wilt disease.] *Transactions of the Meeting of the Japanese Forestry Society.* 86, 287-289.

TANIYAMA, T. (1989). [*Terrible acid rain.*] Tokyo, Japan, Gohdoh Shuppan, 143 pp.

Nematology Monographs & Perspectives, 2003, Vol. 1, 215-223

Characteristics of Japanese red pine (*Pinus densiflora*) families resistant to pine wilt

Keiko KURODA [1] and Hiroyuki KURODA [2]

[1] *Forestry and Forest Products Research Institute, Kansai Research Center, Kyoto 612-0855, Japan*
[2] *Wood Research Institute, Kyoto University, Uji, Kyoto 611-0011, Japan*

Summary – In the trunks of *Pinus* species susceptible to pine wilt disease, blockage of sap ascent becomes extensive after infection by the pine wood nematode (PWN, *Bursaphelenchus xylophilus*) because many tracheids are filled with gas. Infected trees die from a deficit of water supplied to the shoots. In Japan, small numbers of trees of susceptible species, *Pinus densiflora* and *P. thunbergii*, survive in forests that are extensively damaged by pine wilt disease. Trees were selected from many districts and clones are being maintained in seed orchards. Seedlings from selected families express low mortality rates after inoculation with the pathogen. We discuss the factors preventing development of the symptoms in resistant families of *P. densiflora* based on an investigation of the reaction of tree tissues to the infection. In the stems of resistant families, cavitated areas emerged, but those areas did not enlarge. Nematode populations were small in the main stems. There must be systems that prevent nematode migration and reproduction in pine tissue. Numerous branches are a visible characteristic of some resistant families. The complicated arrangement of resin canals may be a physical barrier to PWN migration. Based on the hypothesis that a chemical barrier may be activated in pine tissue, a systemic chemical was tested on non-resistant seedlings. The chemical induced nematicidal stilbenoids in the base of the branches. Symptoms were partially developed or delayed by application of the chemical prior to PWN inoculation. Thus, the base of branches plays an important role in preventing the advance of the nematode.

Japanese black pine (*Pinus densiflora*) and red pine (*P. thunbergii*) are susceptible to pine wilt (Kuroda *et al.*, 1991a, 1998). In the heavily damaged forests of those species, small numbers of trees have been observed to survive even after infection with the pine wood nematode

Table 1. *Resistant families of* Pinus densiflora *inoculated with* Bursaphelenchus xylophilus *and their degree of resistance.*

Age (y)	Name of family	Degree of resistance*	Survival rates (%)**
4 and 5	Matsushima70	3	86-100
4 and 5	Saganoseki165	4	98-100
2	Kamokata29	4	87-95
2	Uwajima50	4	80-100
4 and 5	non-resistant	–	*ca* 20
2	non-resistant	–	*ca* 20

* Kyushu Regional Association of Forest Research Institutions (1999).
** Personal communication from the Forest Tree Breeding Center.

(PWN, *Bursaphelenchus xylophilus*). Such trees were selected from many districts in Japan, and the clones are being maintained in seed orchards in regional offices of the Forest Tree Breeding Center, which annually provides seedlings of the resistant families. The seedlings actually indicate low mortality rate after inoculation with the PWN (Kyushu Regional Association of Forest Research Institutions, 1999; See Table 1, survival rates). However, the factors controlling resistance within a species have not been elucidated.

The PWN invades *via* injuries on young shoots and moves through resin canals in the cortex and xylem (Fig. 1). In the susceptible species, sap ascent is disturbed in the trunks by the initial symptom such as discoloration of old leaves about 3 weeks after infection (Kuroda *et al.*, 1988). Injection of a dye solution into the base of the trunk makes the course of the sap ascent visible. The sap ascends spirally in healthy pine trees. When the sap ascent is disturbed in infected trees, white patches in which the tracheids have been filled with gas emerged in the sapwood. This phenomenon is called cavitation or embolism (Zimmermann, 1983; Kuroda, 1991, 1995, 2000). Cavitated tracheids are never refilled with water in infected trees (Kuroda *et al.*, 1988). When the xylem sap ascent is blocked, trees suffer from a water deficit. The assumed wilting mechanisms are as shown (Fig. 1). A strategy to protect the trees requires the synthesis of secondary products such as terpenoids. However, this is ineffective against PWN and allows the dispersal of the pathogen throughout a tree. Therefore, the dysfunction of sapwood progresses widely in the trunks. Infected trees die when the water supply to the

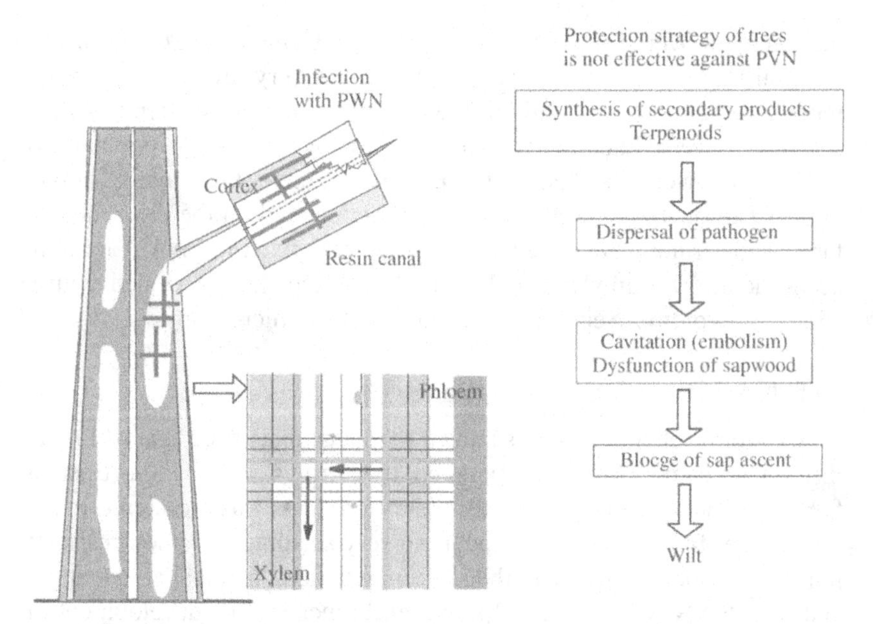

Fig. 1. *Schematic illustration of wilting mechanisms hypothesised for the pine wilt.*

shoots is cut off (Kuroda, 1989, 1991). In highly resistant species, such as *P. taeda*, water blockage was found to occur in the restricted area, but the cavitated areas did not enlarge (Kuroda *et al.*, 1991b).

To clarify the factors related to resistance to the pine wilt disease in some selected families, we investigated the characteristics of the resistant families by focusing on nematode migration and population growth in the pine tissue, and on symptom development. Furthermore, we discuss the structural factors that prevent nematode activities and the chemical factors that can be induced in tree tissues and have possible nematicidal effects.

Materials and methods

NEMATODE POPULATION AND ANATOMY

Inoculation of *B. xylophilus* (S61-strain) incubated on *Botrytis cinerea* was made on four resistant families: Matsushima70, Saganoseki165, Kamokata29, and Uwajima50, as well as on two non-resistant families (Table 1). The survival rates of the resistant families 8 weeks after inoculation were over 80% (Forest Tree Breeding Center, pers. comm.). The survival rate of the non-resistant families was very low, *ca* 20%. After inoculation at the end of July or the beginning of August, samples were taken at 1 week intervals for about 2 months. Nematode populations in the stems were checked by the Baermann funnel technique (Thorne, 1961). The degree of water blockage was detected on the crosscut surface of the trunks. Dissected tree tissues were fixed in FAA (formalin, acetic acid, 50% ethyl alcohol; 5:5:90 v/v) and then sectioned with a sliding microtome. Sections were processed for microscopy.

APPLICATION OF A SYSTEMIC CHEMICAL

A systemic chemical that is known to induce resistance against fungal disease in some plants was applied to cuttings of the non-resistant families. The concentration of secondary products was monitored in the pine tissue. In a second experiment, nursery seedlings of a non-resistant family were pre-treated with the abovementioned chemical at irrigation, and then PWN was inoculated into their branches. The development of symptoms was monitored.

Results and discussion

POPULATION GROWTH

The PWN population increases in the stems of Matsushima70, Saganoseki165, and a non-resistant family are shown in Fig. 2. In all families, nematode population was large in the inoculated branches. Differences were found in the distribution in the main stems. In the non-resistant family, nematodes swiftly migrated to the base of the main stems and increased for 6 weeks. In resistant families, the distribution of the nematodes was slow in the stem for 6 weeks after the inoculation. The population was very small in the main stems at 4 weeks after infection, except in some specimens. It was clear that population growth was retarded or prevented in the stems of the resistant families. There must be systems that prevent nematode migration and reproduction in pine tissue.

In the case of two other resistant families, Uwajima50 and Kamokata29, population growth was observed in some specimens. The younger seedlings of these families were not as resistant as 4-year-old Matsushima70 and Saganoseki165. In these cases, population growth of PWN in the roots was slower.

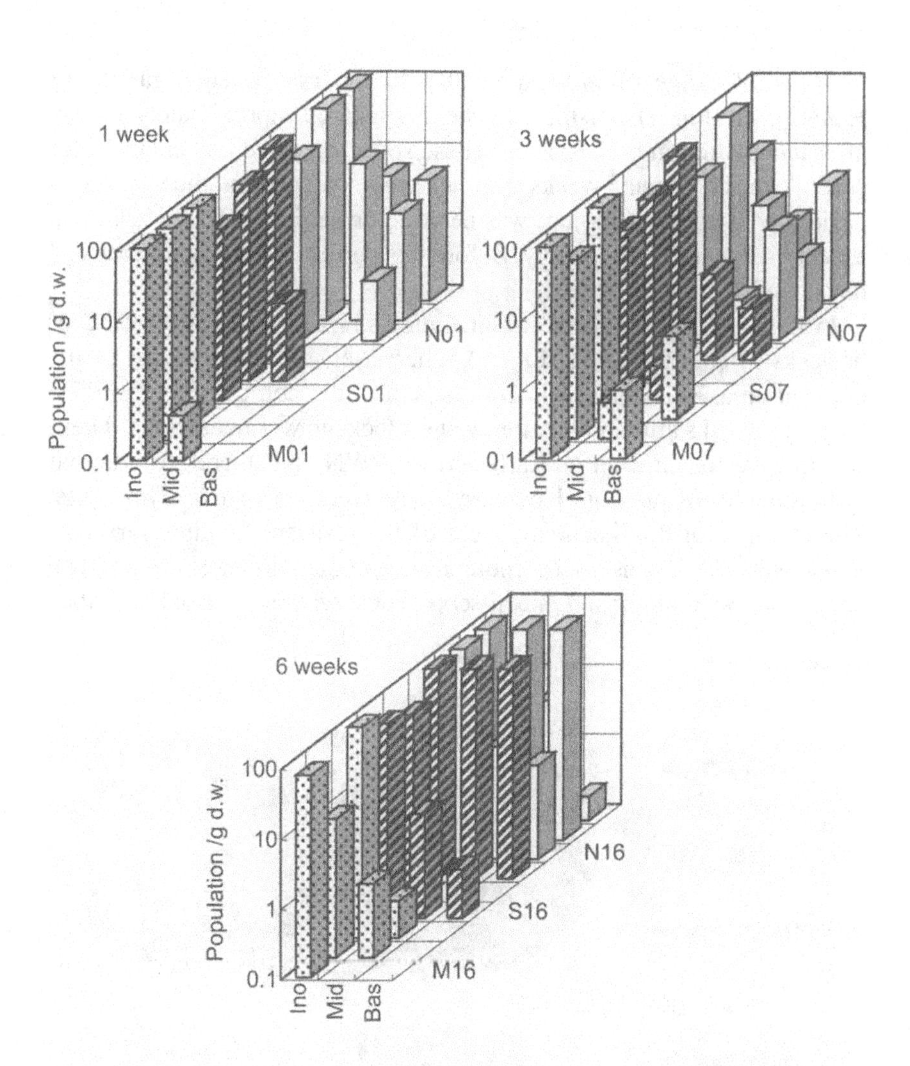

Fig. 2. *Population growth of* Bursaphelenchus xylophilus *in the inoculated branches and stems of three* Pinus densiflora *families at 1, 3 and 6 weeks after inoculation. Ino: Inoculated branch; Mid: Middle of main stem; Bas: base of main stem.*

WATER BLOCKAGE

Water blockage progressed swiftly in the non-resistant family of *P. densiflora* (Fig. 3). Air-filled tracheids appeared approximately 1 week after inoculation in both species. These spots increased and enlarged for several weeks. About 5 weeks after nematode inoculation, the xylem was desiccated, and the cambium was necrotic or partially necrotic. The sap ascent had stopped completely. Before this period, visible symptoms had developed.

In the seedlings of the resistant families, small spots appeared 2 or 3 weeks after inoculation (Fig. 3). Thereafter, the enlargement of the affected areas was delayed in comparison to the non-resistant families. Even at 5 weeks after inoculation, water blockage was not serious. These seedlings were infected and affected by PWN and, therefore, internal symptoms were present. However, there were no visible symptoms. The majority of the inoculated trees of the resistant families survived. Saganoseki165 seems to be more resistant than Matsushima70. The nematode population and macroscopic observation of trunk sections

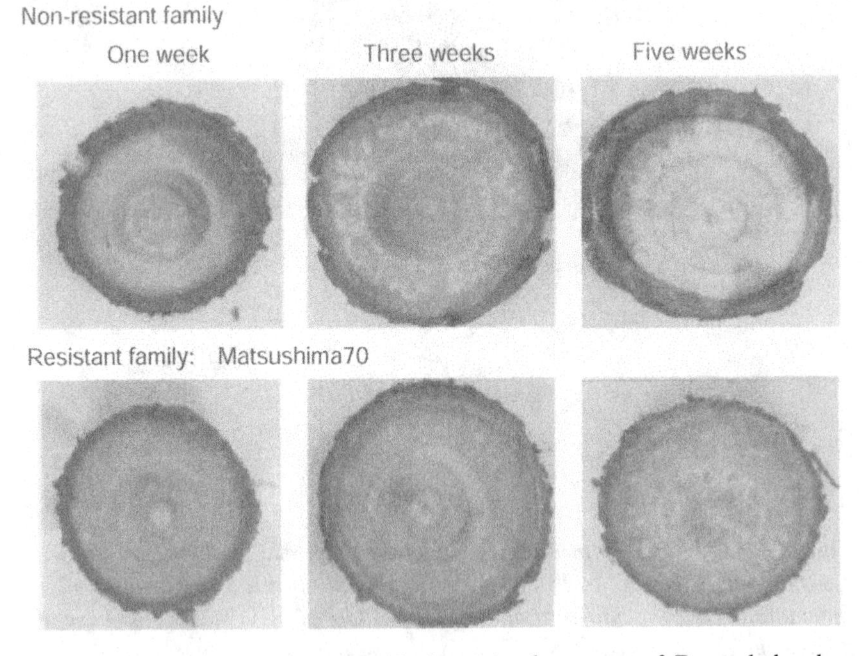

Fig. 3. *Development of water blockage in the stems of* Bursaphelenchus xylophilus *resistant and non-resistant families of* Pinus densiflora.

suggest that blockage of the sap ascent did not develop due to the weak activities of PWN in the resistant seedlings.

ANATOMICAL AND CHEMICAL FACTORS PREVENTING NEMATODE ACTIVITIES

A hypothesis was developed based on the factors preventing PWN activities. An anatomical factor may prevent nematode distribution. Numerous branches and big knots are characteristic of some resistant families (Fig. 4). The non-resistant seedlings had approximately five branches at the base of inoculated branches. Saganoseki165 and Matsushima70 had approximately eight or more branches. In Matsushima70 (Fig. 4B), the nematode population in the main stem was extremely small, even at 7 weeks after inoculation. Radial sections of the knots indicated a disorder in the arrangement of the tissue. Tracheids and resin canals were winding and crossing at the base of the branches. PWN uses resin canals for dispersal and would have difficulty traversing the disordered resin canals to enter the trunk. The structure acts as a physical barrier to nematode migration.

The resistant families may activate a chemical barrier in their tissues after PWN infection. The induction of resistance was tried by the application of a systemic chemical to the cuttings of non-resistant

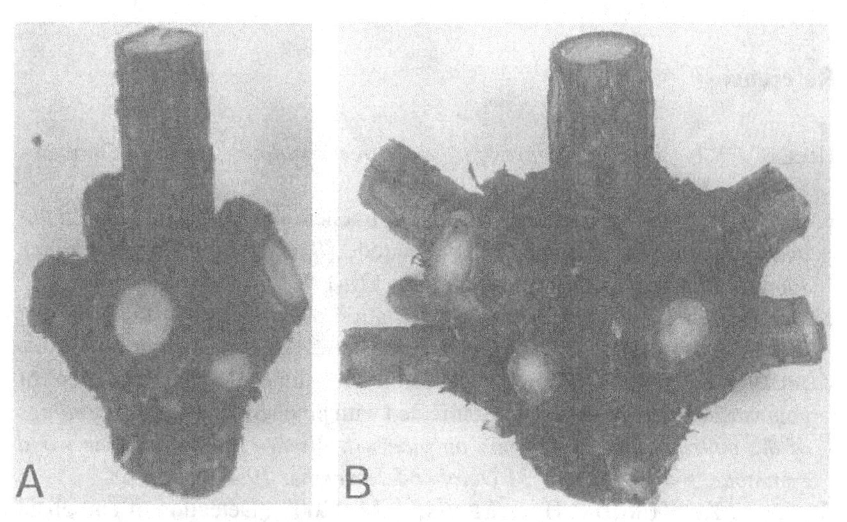

Fig. 4. *Branch morphology of two families of* Pinus densiflora. *A: Non-resistant; B: Resistant Matsushima70.*

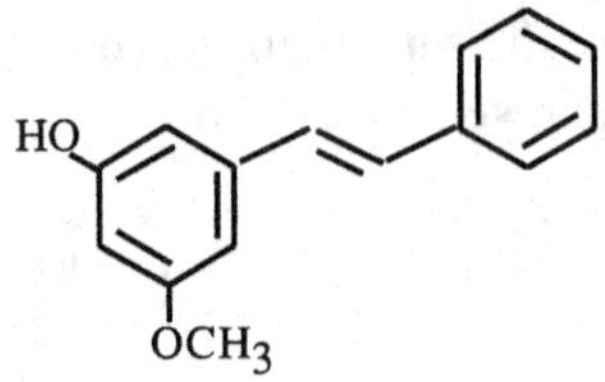

Fig. 5. *Molecular structure of the stilbenoid, pinosylvin monomethylether (PMM).*

families. Stilbenoids, especially pinosylvin monomethylether (PMM), were induced (Fig. 5). This phenomenon occurred at the branch bases but not in the shoots. Stilbenoids are secondary products synthesised at heartwood formation, and they do not increase in sapwood (Hillis, 1987). In this experiment, the chemical switched the enzymatic activities on stilbenoid synthesis in the sapwood. PMM is known to have nematicidal effects. Treatment of non-resistant seedlings with the chemical and inoculation with PWN induced the death of inoculated branches only, and, in some cases, retarded wilting. The nematode effects were clearly prevented at the base of the branches. Furthermore, in this case, the bases of the branches played an important role. In this experiment, we found that stilbenoid formation is one of the key strategies for preventing pine wilt.

References

HILLIS, W.E. (1987). *Heartwood and tree exudates.* Berlin, Germany, Springer-Verlag, 268 pp.

KURODA, K. (1989). Terpenoids causing tracheid-cavitation in *Pinus thunbergii* infected by the pine wood nematode (*Bursaphelenchus xylophilus*). *Japanese Journal of Phytopathology* 55, 170-178.

KURODA, K. (1991). Mechanism of cavitation development in the pine wilt disease. *European Journal of Forest Pathology* 21, 82-89.

KURODA, K. (1995). Acoustic emission technique for the detection of abnormal cavitation in pine trees infected with pine wilt disease. *Proceedings of the international symposium on pine wilt disease caused by pine wood nematode, Beijing, China, 31 October-5 November 1995*, pp. 53-58.

KURODA, K., KURODA, H. & LEWIS, A.M. (2000). Detection of embolism and acoustic emissions in tracheids under a microscope: incidence of diseased trees infected with pine wilt. In: Kim Y.S. (Ed.). *New Horizons in Wood Anatomy*. Kwangju, Korea, Chonnam National University Press, pp. 372-377.

KURODA, K., YAMADA, T. & ITO, S. (1991a). [Cavitation development in *Pinus densiflora* studied from a stand point of water conduction.] *Journal of the Japanese Forestry Society* 73, 69-72.

KURODA, K., YAMADA, T. & ITO, S. (1991b). *Bursaphelenchus xylophilus* induced pine wilt: factors associated with resistance. *European Journal of Forest Pathology* 21, 430-438.

KURODA, K., YAMADA, T., MINEO, K. & TAMURA, H. (1988). Effects of cavitation on the development of pine wilt disease caused by *Bursaphelenchus xylophilus*. *Japanese Journal of Phytopathology* 54, 606-615.

KYUSHU REGIONAL ASSOCIATION OF FOREST RESEARCH INSTITUTIONS (1999). [Characteristics of Japanese cypress and resistant families of pine trees.] *Publication of the Breeding Working Party*, Kyushu, Japan, 58 pp.

THORNE, G. (1961). *Principles of nematology*. New York, NY, USA, McGraw-Hill, pp. 48-49.

ZIMMERMANN, M.H. (1983). *Xylem structure and the ascent of sap*. Berlin, Germany, Springer-Verlag, 143 pp.

Nematology Monographs & Perspectives, 2003, Vol. 1, 225

A study on variation of pH between healthy and pine wilt diseased woods of several pine species

Yuyan WANG, L. HAIYAN, S. CHAORAN and G. ZHIHONG

General Station of Forest Pest & Disease Management, Shenyang 110034, China

Summary – An analysis of the variation of pH between healthy and pine wilt diseased woods including Japanese black pine (*Pinus thunbergii*), Japanese red pine (*P. densiflora*), masson pine (*P. massoniana*), loblolly pine (*P. taeda*) and slash pine (*P. elliottii*) from two epidemic areas in Jiangsu and Anhui has been completed. The differences in pH between healthy and diseased woods of Japanese black pine in Jiangsu, Anhui and Japanese red pine in Jiangsu were extremely notable. The difference in pH between healthy and diseased woods of masson pine in Anhui was also extremely notable. The difference in pH between healthy and diseased woods of masson pine in Jiangsu was not obvious. The difference in pH between healthy and diseased woods of loblolly pine was significant, but not obvious for slash pine. Thus it can be seen that the difference in pH between healthy and diseased woods varied in pines, and the differences were consistent in two provinces. The more susceptible and more severe the disease, the more obvious was the difference in pH. The difference in pH may be used in the rapid quarantine inspection for pine wilt disease.

Nematology Monographs & Perspectives, 2003, Vol. 1, 227-237

Pine wood nematode phoresis: the impact on *Monochamus carolinensis* life functions

Marc LINIT [1] and Suleyman AKBULUT [2]

[1] *Department of Entomology, University of Missouri, Columbia, Missouri 65211, USA*
[2] *Forest Entomology and Protection Unit, Department of Forestry, Abant Izzet Baysal University 81150, Duzce, Turkey*

Summary – The phoretic relationship between *Monochamus carolinensis* (Olivier) and the pine wood nematode, *Bursaphelenchus xylophilus* (Steiner & Buhrer, 1934) Nickle, 1970, is obligatory for the nematode but poorly understood for the beetle. This relationship is important for a complete understanding of the interspecific association of these symbiotic organisms and how that association affects beetle reproduction and population dynamics. To better understand symbiosis between the nematode and its insect vector we studied the relationship between beetle flight performance and the number of fourth-stage dispersal juvenile (J4) pine wood nematodes carried by individual *M. carolinensis* adults (nematode load) and the influence of nematode load on beetle reproductive potential. The data presented suggest that the reproductive potential of the majority of beetles in a population is unaffected by the number of J4 carried, resulting in a commensal relationship under endemic conditions that is obligatory for the nematode and facultative for the vector.

The phoretic relationship between *Monochamus carolinensis* (Olivier) and the pine wood nematode, *Bursaphelenchus xylophilus* (Steiner & Buhrer, 1934) Nickle, 1970, is obligatory for the nematode but poorly understood for the beetle. Individual adult *M. carolinensis* commonly carry up to 20 000 fourth stage juveniles (J4) upon emergence from nematode-infested trees. Occasionally, individual beetles carry more than 50 000 J4 (Linit, 1988). The presence of many J4 in the tracheae of *Monochamus* beetles may negatively impact on life processes of the vector. Togashi and Sekizuka (1982) documented the negative impact of nematode phoresy on longevity of *M. alternatus* Hope, while Humphry and Linit (1989a) noted a negative impact on flight duration of *M. carolinensis*. The impact of nematode load (the number of J4 carried per beetle) on important life functions of its beetle vector is poorly understood. This relationship is important for a complete understanding of the interspecific association of these symbiotic organisms and how that association affects beetle reproduction and population dynamics. Flight is an intensive, energy-demanding insect activity, and its maintenance is dependent upon a sufficient and continuous supply of fuel and oxygen to the flight muscles. Oxygen is supplied to the flight muscles by metathoracic tracheoles, therefore tracheal packing by nematodes might influence the flight performance of *M. carolinensis*. The intrinsic rate of increase (r), a measure of the rate of population growth, is the rate of increase per head under specific physical conditions in an unlimited environment (Birch, 1948). This parameter can be used to compare the reproductive potential of different species or populations of a single species under different environmental or physiological conditions. To better understand symbiosis between the nematode and its insect vector we studied the relationship between beetle flight performance and the number of fourth-stage dispersal juvenile (J4) pine wood nematodes carried by individual *M. carolinensis* adults (nematode load) and the influence of nematode load on the beetle's reproductive potential.

Materials and methods

FLIGHT PERFORMANCE

Beetles, newly emerged from jack pine (*Pinus banksiana* Lamb.) logs reared under laboratory conditions, were collected daily, weighed to the nearest mg, and marked on the elytra according to the method of Humphry and Linit (1989b) to maintain their identity. The beetles were examined within 1 day after emergence to categorise the number of J4 carried using the method of Zhang *et al.* (1995). Beetles carrying between zero and 100 J4 were considered non-infested for the purposes of this study.

A flight mill was constructed to evaluate beetle flight performance (Fig. 1). The beetle was attached to the flight arm by inserting the sharp end of the insect pin into a cork near the distal end of the flight arm. The beetle was stimulated to initiate flight by removing tarsal contact with a solid surface or by passing a stream of air over its body. If flight was not initiated within 10 s, the beetle was removed from the flight mill.

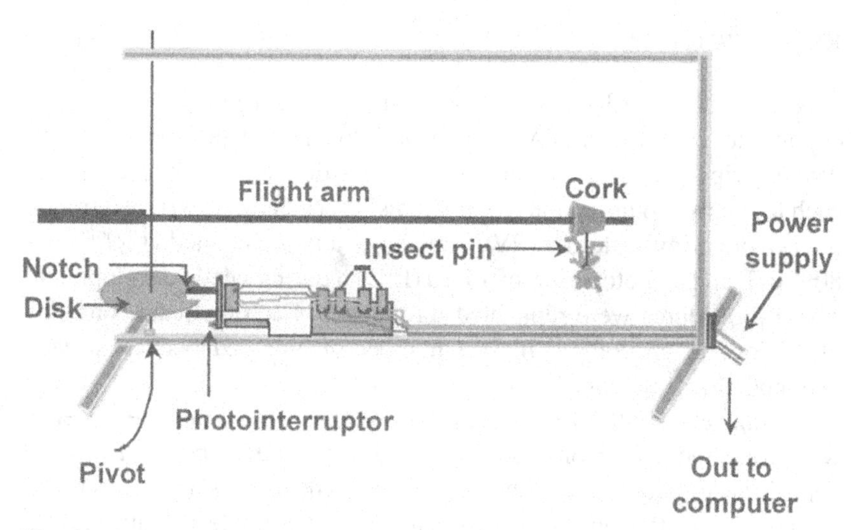

Fig. 1. *Flight mill constructed to evaluate beetle flight performance.*

The number of revolutions and the time of each revolution were determined by a General Electric H11A1 photointerruptor, consisting of an infrared LED and a phototransistor that was energised as a notch in a disk connected to the flight arm passed the cell during each revolution (Fig. 1). A BASIC program was written to record the number of revolutions for each 15 s time period from start through to the termination of beetle flight. A single flight from start to finish with no interruptions was recorded for each beetle used in the study. Beetles within 24 h of emergence from logs and prior to any feeding were used to monitor flight performance. After cessation of flight, the beetle was macerated and nematodes were collected for 24 h using the Baermann funnel technique (Southey, 1986). A detailed description of this process can be found in Akbulut and Linit (1999a).

The distribution of J4 carried by individuals within a population of *Monochamus* adults is contagious with the majority of beetles carrying fewer than 10 000 J4 and a few beetles carrying up to 50 000 and rarely 100 000 J4 (Linit, 1988). Thus, the beetle population was divided into two categories based on the number of J4 recovered from each beetle: *i*) beetles carrying less than 10 000 and *ii*) beetles carrying more than 10 000 J4.

REPRODUCTIVE POTENTIAL

Newly cut logs (38-41 cm long, 12-14 cm diam.) of jack pine were exposed to a lab colony of *M. carolinensis* to allow oviposition. The start date of oviposition and the number of oviposition sites were recorded for each log. After counting the oviposition sites, each log was transferred to its own polyvinyl chloride (PVC) container and maintained at 30°C, 70-80% RH, and a photoperiod of 14:10 (L:D) h during beetle development. These procedures were replicated three times. Each replicate consisted of 13 logs to ensure a sufficient number of emerged beetles in each nematode load category.

Starting about 60-70 days after oviposition, newly emerged beetles were collected daily from each log and the age of each beetle, from egg deposition to adult emergence, was noted. The beetles were examined within 1 day after emergence to categorise the number of nematodes carried using the non-destructive method of Zhang *et al.* (1995). Each beetle was placed in one of three nematode-load categories, low (0-100 J4), medium (101-20 000 J4) or high (>20 000 J4) according to this method. Each nematode load category constituted a treatment group for the comparison of population parameters.

Each emerged female beetle was maintained in its own paper container fitted with screen lids and maintained at 30°C, 70-80% RH, and a photoperiod of 14:10 (L:D) h during the experiment. Male beetles were rotated among their female partners on a 5 day basis to eliminate differences in sperm quality or quantity on a female ovipositional performance. Each pair was provided with fresh pine twigs for feeding, which were replaced every 48 h, and small pine bolts for oviposition. The total number of eggs oviposited by each female during her life span was determined by examination of the oviposition bolts.

Several population parameters were determined for beetles in each nematode load category based upon survival and reproduction of females (Birch, 1948). The time interval (x) for time dependent functions was 5 days. The sex ratio of eggs was assumed to be 0.60 (female/total number of beetles) based on the ratio of adults emerged during this study. Initial female egg density and the within-wood (egg to adult female emergence) survival rate of immature stages, were estimated. The initial egg density for each log was estimated by multiplying the number of oviposition sites by 1.014. This constant was determined in a preliminary study in which 566 oviposition sites were dissected and the number of eggs

Table 1. *Mean (±SE) weight, number of fourth stage juveniles of pine wood nematode (J4) recovered per beetle, flight distance, duration, and speed for 1-day-old* Monochamus carolinensis *adult females and males.*

Beetle sex	Number of beetles	Variable[a]				
		Adult weight (mg)	Number of J4 recovered	Flight distance (m)	Flight duration (min)	Flight speed (m/min)
Female	64	308 ± 0.05 a	2465 ± 5417.1 a	2264 ± 1121.1 a	25.3 ± 12.43 a	89.8 ± 11.46 a
Male	53	314 ± 0.06 a	3656 ± 7936.3 a	2122 ± 1113.9 a	23.4 ± 11.60 a	89.5 ± 11.59 a
Pooled	117	311 ± 0.05	3004 ± 6672.8	2200 ± 1115.3	24.4 ± 12.04	89.7 ± 11.47

[a] Means within each column followed by the same letter did not differ according to Fisher's Least Significant Difference ($P = 0.05$). Statistical analysis of the nematode data was conducted on square-root transformed data, non-transformed means for variable J4 are reported above.

determined for each. A detailed description of the process can be found in Akbulut and Linit (1999b).

A fertility table (Birch, 1948) was constructed by collecting age-specific survivorship and fecundity data. The age-specific survival rate of adults (l_x), the proportion of females alive at time x and the age-specific fecundity (m_x), the expected number of female progeny per female at age x, were calculated for each time interval ($x = 5$ days). In addition, adult female longevity and time-specific egg deposition rates were determined for each nematode load category.

Results

Flight performance

Male beetles carried a greater mean number of J4 than females, but variation was high and the difference was not significant ($F = 0.42$; df = 1, 115; $P = 0.516$) (Table 1). Likewise, mean beetle weight did not differ ($F = 0.35$; df = 1, 115; $P = 0.553$) between the sexes. Flight distance ($F = 0.47$; df = 1, 115; $P = 0.494$), duration ($F = 0.67$; df = 1, 115; $P = 0.417$) and velocity ($F = 0.03$; df = 1, 115; $P = 0.856$) were very similar between sexes.

Table 2. *Mean (±SE) weight, number of J4 recovered per beetle, flight distance, duration, and speed for 1-day-old* Monochamus carolinensis *adult beetles carrying <10 000 or >10 000* Bursaphelenchus xylophilus *J4.*

Number of J4	Number of beetles	Variable[a]				
		Adult weight (mg)	No J4 recovered	Flight distance (m)	Flight duration (min)	Flight speed (m/min)
<10 000	106	310 ± 0.05 a	1123 ± 1954.0 a	2274 ± 1112.1 b	25.4 ± 12.01 b	89.4 ± 11.48 a
>10 000	11	316 ± 0.05 a	21 133 ± 8796.8 b	1484 ± 905.9 a	15.7 ± 8.62 a	92.58 ± 11.57 a

[a] Means within each column followed by the same letter did not differ according to Fisher's Least Significant Difference ($P = 0.05$). Statistical analysis of the nematode data was conducted on square-root transformed data, non-transformed means for variable J4 are reported above.

The relationships between J4 load and flight duration, distance, and velocity were not significant for either sex or for the pooled data when analysed by linear regression. However, when beetles were categorised by nematode load, differences in flight performance were observed. Beetles with a low nematode load (<10 000) flew significantly further ($F = 5.18$; df = 1, 115; $P = 0.025$) and for a longer duration ($F = 6.67$; df = 1, 115; $P = 0.011$) than beetles with a high nematode load (>10 000). Mean flight velocity did not differ ($F = 0.73$; df = 1, 115; $P = 0.396$) (Table 2).

REPRODUCTIVE POTENTIAL

The age-specific pattern of egg deposition was similar for females in all nematode load categories through the first half of the oviposition period (Fig. 2A). The duration of the oviposition period was inversely related to nematode load. Beetles with a low nematode load continued oviposition activity through a greater age than beetles with a high load. The total number of eggs deposited by beetles (*i.e.*, fecundity), however, did not differ significantly ($F = 0.43$; df = 2, 97; $P = 0.654$) among nematode load categories (Table 3).

The longevity of adult females was not affected by nematode load ($F = 1.31$; df = 2, 97; $P = 0.274$). However, because individuals with rapid within-wood development tended to carry more J4 than those that

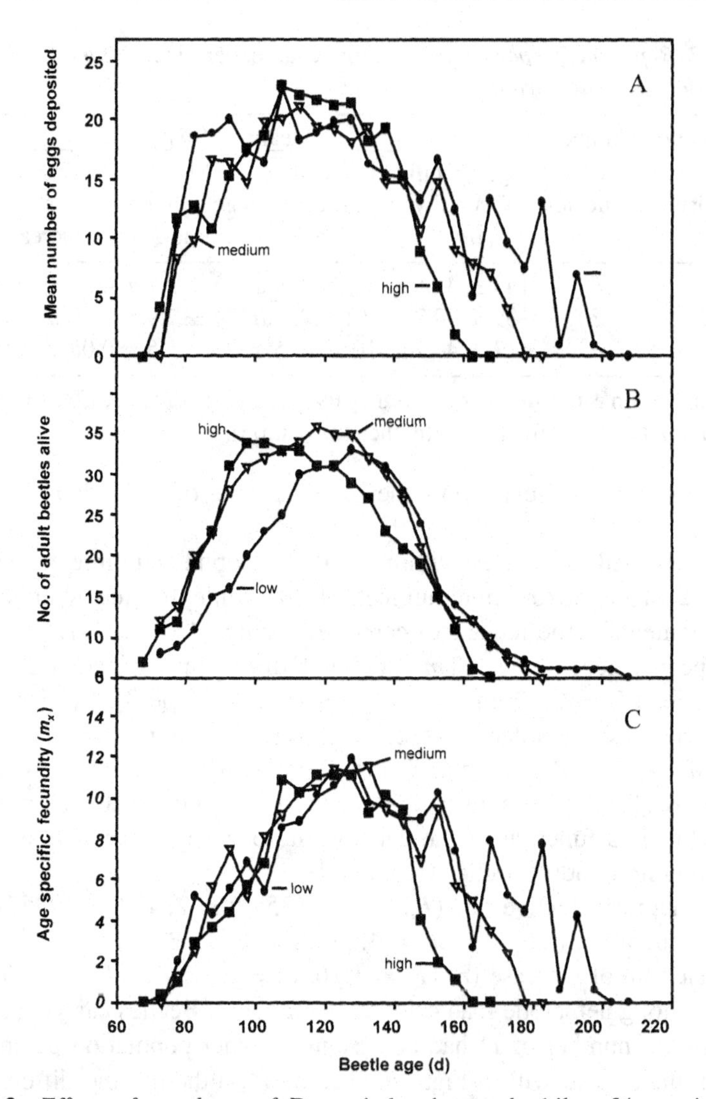

Fig. 2. *Effect of numbers of* Bursaphelenchus xylophilus *J4 carried on* Monochamus carolinensis *females. A: Age specific pattern of egg deposition; B: Number of living adult females; C: Age specific fecundity (*m_x*). Values designating beetle age are the mid-points of each interval; ●, ▽, ■: low, medium and high numbers of J4.*

developed slowly, total longevity (egg to adult death) was influenced ($F = 3.76$; df = 2, 97; $P = 0.027$) by nematode load. There was a significant ($b = 0.217$; $SE_b = 0.008$; $P < 0.001$) linear relationship

Table 3. *Reproductive parameters of* Monochamus carolinensis *females in three nematode load categories.*

Nematode load category	Number of replicates	Mean[a] (±SE): Age-specific fecundity im_x	Next reproductive rate R_o	Mean generation time T	Intrinsic rate of increase r
Low	3	143 ± 35.5 a	12 ± 10.9 a	25 ± 2.1 a	0.0930 ± 0.05 a
Medium	3	139 ± 33.7 a	12 ± 9.1 a	24 ± 1.9 a	0.0962 ± 0.04 a
High	3	112 ± 44.2 a	10 ± 10.5 a	23 ± 1.5 a	0.0852 ± 0.05 a

[a] Means within each column followed by the same letter did not differ according to Fisher's Least Significant Difference ($P = 0.05$).

between adult longevity and the total number of eggs deposited by beetles.

Beetles with a shorter within-wood developmental time tended to carry a moderate or high number of J4, while beetles with longer developmental time tended to carry few or no J4 (Fig. 2B). The total age-specific fecundities ($\%m_x$) did not differ among nematode load categories ($F = 2.78$; df = 2, 4; $P = 0.175$) (Table 3). Beetles in all categories had a similar age-specific fecundity pattern through the first half of the fecundity schedule (Fig. 2c). The duration of age-specific fecundity schedule was inversely related to nematode load. Age-specific fecundity is a function of the age-specific egg deposition rate and the survivorship of adult beetles (Fig. 2A, B).

The net reproductive rate (R_o) ($F = 1.35$; df = 2, 4; $P = 0.357$), the mean generation time (T) ($F = 4.09$; df = 2, 4; $P = 0.108$), and the intrinsic rate of increase (r) ($F = 1.16$; df = 2, 4; $P = 0.401$) did not differ among nematode load categories (Table 3). Beetles carrying a low or medium number of J4 had consistently higher population parameter values than those with a high number of J4, although the differences were never significant.

Discussion

Most beetles within a population carry few J4 upon their emergence from host trees, while a few beetles carry many (Kobayashi *et al.*, 1984; Linit, 1988). Thus, the nematode population that exits an infested tree is aggregated within a small percentage of emerging beetles. In the present study, the flight performance of beetles that carried a high number (>10 000) of J4 was reduced compared to beetles that carried fewer J4. Consequently, the impact of nematode load on the flight performance of a population of *M. carolinensis* is limited to a few beetles that carry a high number of J4. Flight capacity was reduced by 38% in these beetles; however, the beetles still had a mean continuous flight of more than 15 min. This suggests that the vital capacity of the *M. carolinensis* respiratory system is not severely limited by pine wood nematode infestation.

Overall, nematode load did not have a significant negative effect on flight performance of *M. carolinensis* except for beetles that carried a high number of nematodes. The mated status of female beetles significantly affected flight capabilities. In Japan, pine wilt disease is epidemic (Kishi, 1995) and susceptible pine trees are abundant, requiring only short dispersal flights for nematode-infested beetles to locate feeding and oviposition sites. Thus, the effect of nematode load or mated status on the flight performance of *M. alternatus* may not be a critical factor in the epidemiology of pine wilt. In North America, pine wilt is primarily associated with exotic, ornamental pines. The limited and widely separated distribution of these trees may require a long flight by *M. carolinensis* to locate feeding and oviposition sites. Thus, nematode load or mated status may negatively influence the ability of a beetle to locate food and oviposition sites.

The intrinsic rate of increase is influenced by several population processes, including age-specific survivorship, age-specific fecundity and the developmental time until first reproduction (Birch, 1948). In the present study, significant differences in the intrinsic rates of increase among beetles in three nematode load categories were not detected. Differences in the length of the within-wood development period, age-specific survivorship and patterns of oviposition among beetle groups, however, resulted in small differences in this parameter.

Birch (1948) documented that early egg deposition makes an important contribution to the value of r. In order to have equivalent values of r, a population of organisms with slow immature development would need to have greater total fecundity than a similar population that reached reproductive age in less time. Zhang and Linit (1998) reported that the oviposition rate of adult *M. carolinensis* was not age-dependent and total fecundity was a linear function of adult longevity. The oviposition patterns of *M. carolinensis* in the present study fit those observations.

Oviposition began for each cohort at approximately the same age, about 70 days. The duration of the oviposition period was longest for beetles with low numbers of J4 and shortest for those with the greatest number. This extended period of oviposition did not significantly influence the intrinsic rates of increase because beetle survivorship beyond the age of 160 days was so low that the contribution to $l_x m_x$ was negligible.

The intrinsic rate of increase provides a means to compare the impact of an environmental constraint on the survivorship and reproduction of a population of organisms. In the present study, the constraint studied was pine wood nematode phoresis on the vector, *M. carolinensis*. Beetles that carried low or medium nematode loads had rates of increase that were very similar. Beetles that carried a high number of J4 had a rate that was slightly, but not significantly, lower than the other cohorts. This suggests that the survival and reproduction of beetles that carry fewer than 20 000 nematodes are unaffected by pine wood nematode phoresy.

Several studies have reported individual vectors of the pine wood nematode that have carried more than 50 000 and occasionally more than 100 000 J4 (Kobayashi *et al.*, 1984; Linit, 1988). While extremely high nematode loads are uncommon, it is not unusual for individuals within a cohort of beetles to carry more than 50 000 J4. Had we been able to segregate a cohort of beetles with extremely high nematode loads (>50 000), we expect a significant reduction in the intrinsic rate of increase, compared to beetles carrying fewer J4, would have been documented. The data presented here, however, suggest that the reproductive potential of the majority of beetles in a population is unaffected by the number of J4 carried resulting in a commensal relationship under endemic conditions that is obligatory for the nematode and facultative for the vector.

References

AKBULUT, S. & LINIT, M.J. (1999a). Flight performance of *Monochamus carolinensis* (Coleoptera: Cerambycidae) with respect to nematode phoresis and beetle characteristics. *Environmental Entomology* 28, 1014-1020.

AKBULUT, S. & LINIT, M.J. (1999b). Reproductive potential of *Monochamus carolinensis* (Coleoptera: Cerambycidae) with respect to pinewood nematode phoresis. *Environmental Entomology* 28, 407-411.

BIRCH, L.C. (1948). The intrinsic rate of natural increase of an insect population. *Journal of Animal Ecology* 17, 15-26.

HUMPHRY, S.J. & LINIT, M.J. (1989a). Effect of pinewood nematode density on tethered flight of *Monochamus carolinensis* (Coleoptera: Cerambycidae). *Environmental Entomology* 18, 670-673.

HUMPHRY, S.J. & LINIT, M.J. (1989b). Tethered flight of *Monochamus carolinensis* (Coleoptera: Cerambycidae) with respect to beetle age and sex. *Environmental Entomology* 18, 124-126.

KISHI, Y. (1995). The pinewood nematode and the Japanese pine sawyer. *Forest Pest in Japan. No. 1.* Tokyo, Japan, Thomas Company Limited, 302 pp.

KOBAYASHI, F., YAMANE, A. & IKEDA, T. (1984). The Japanese pine sawyer beetle as a vector of pine wilt disease. *Annual Review of Entomology* 29, 115-135.

LINIT, M.J. (1988). Nematode-vector relationships in the pine wilt system. *Journal of Nematology* 20, 227-235.

NICKLE, W.R. (1970). A taxonomic review of the genera of the Aphelenchoidea (Fuchs, 1937) Thorn, 1949 (Nematoda: Tylenchida). *Journal of Nematology* 2, 375-392.

SOUTHEY, J.F. (ED.) (1986). *Laboratory methods for work with plant and soil nematodes*. Ministry of Agriculture, Fisheries, and Food, Great Britain, Ref. Book 402, London, UK, Her Majesty's Stationery Office, 202 pp.

STEINER, G. & BUHRER, E.M. (1934). *Aphelenchoides xylophilus*, n. sp., a nematode associated with blue-stain and other fungi in timber. *Journal of Agricultural Research* 33, 159-164.

TOGASHI, K. & SEKIZUKA, H. (1982). Influence of the pine wood nematode, *Bursaphelenchus lignicolus* (Nematoda: Aphelenchoididea), on longevity of its vector, *Monochamus alternatus* (Coleoptera: Cerambycidae). *Applied Entomology and Zoology* 17, 160-165.

ZHANG, X., STAMPS, W.T. & LINIT, M.J. (1995). A nondestructive method of determining *Bursaphelenchus xylophilus* infestation of *Monochamus* spp. vectors. *Journal of Nematology* 27, 36-41.

ZHANG, X. & LINIT, M.J. (1998). Comparison of oviposition and longevity of *Monochamus alternatus* and *M. carolinensis* (Coleoptera: Cerambycidae) under laboratory conditions. *Environmental Entomology* 27, 885-891.

Nematology Monographs & Perspectives, 2003, Vol. 1, 239-245

Obstruction by *Bursaphelenchus mucronatus* of *B. xylophilus* boarding *Monochamus alternatus*

Shota JIKUMARU [1] and Katsumi TOGASHI [2]

[1] *Hiroshima Prefectural Forestry Research Centre, Miyoshi, Hiroshima 728-0015, Japan*
[2] *Department of Environmental Sciences, Faculty of Integrated Arts and Sciences, Hiroshima University, Higashi-Hiroshima, Hiroshima 739-8521, Japan*

Summary – To clarify interspecific interference between *Bursaphelenchus xylophilus* and *B. mucronatus* in the number of nematodes carried by adult *Monochamus alternatus* (nematode load), post-diapause larvae were placed individually in artificial pupal chambers of *Pinus densiflora* bolts inoculated with *B. xylophilus* alone, *B. mucronatus* alone, or both nematode species combined at 25°C. In the case of pine bolts inoculated with a single nematode species, *B. xylophilus* load was significantly greater than *B. mucronatus* load, although there was no difference in the abundance of dispersal juveniles around the pupal chamber between the two treatments. Beetles that emerged from pine bolts inoculated with both nematode species showed lower mean nematode load than those from bolts inoculated with *B. xylophilus* alone, but greater than those from bolts inoculated with *B. mucronatus* alone. The results showed an extremely low ability of *B. mucronatus* to board the beetle compared with *B. xylophilus*, and an inhibitory effect of *B. mucronatus* on the number of *B. xylophilus* carried by beetle.

Bursaphelenchus mucronatus Mamiya & Enda is a closely related species of *B. xylophilus* (Steiner & Buhrer) Nickle, the causative agent of pine wilt disease. *Bursaphelenchus mucronatus* is native to Japan and avirulent against native species of the genus *Pinus* in Japan (Mamiya & Enda, 1979). In contrast, *B. xylophilus* is native to North America and believed to have been introduced into Japan in the early 1900s (Rutherford & Webster, 1987). It is virulent against many of native *Pinus* species in eastern Asia and has been causing devastating damage to the pine forests (Mamiya, 1988).

The two nematode species are practically identical in their biology. Thus, a rigorous interspecific interaction is expected between the two nematode species when they are distributed sympatrically. In fact, after the invasion of Japan by *B. xylophilus, B. mucronatus* appears to have been replaced in the pine forests with pine wilt disease (Kishi, 1995). However, the mechanism of the species replacement has not yet been clarified, although there have been a few articles on interaction between two nematode species. Wood-boring beetles provide the only known means of transport for these nematodes from infested to uninfested host trees (Linit, 1988). Thus boarding ability is crucial for the nematodes to persist in forest ecosystems, and interspecific competition might be severe at the time of boarding when the two nematode species exist in a tree. In this paper, we compare the boarding ability of the two nematode species by using the vector, *Monochamus alternatus* Hope, a primary vector of *B. xylophilus* in Japan, and evaluate the degree of interspecific competition at the time of boarding.

Materials and methods

Bursaphelenchus mucronatus and *B. xylophilus* were cultured monoxenically on *Botrytis cinerea* Pers. at 25°C. The origin of the nematodes was Takano Town, Hiroshima Prefecture. After multiplication on fresh *B. cinerea* for 2-3 weeks, they were extracted at 25°C for 24 h using the Baermann funnel technique and adjusted to a suspension of 3000 nematodes including juveniles and adults per 1 ml of distilled water. A suspension of two combined nematode species was made by mingling equal quantities of two suspensions of different nematodes.

The method developed by Aikawa and Togashi (1997) was used for loading *M. alternatus* adults with nematodes. Bolts were made of healthy *P. densiflora* Sieb. & Zucc. trees at Hiroshima Prefectural Forestry Research Centre, Miyoshi City, Hiroshima Prefecture. Mean length and diameter of the bolts were *ca* 7 and 4 cm, respectively. A hole (*ca* 5 cm in depth and *ca* 1 cm diam.) was drilled in the centre of a cut end of each bolt as an artificial pupal chamber. The bolts were set upright individually on quartz sand in polycarbonate containers (11 cm high and 6.7 cm inside diam., Agripot®, Kirin Co.). They were autoclaved at 121°C for 15 min, and then each bolt was inoculated with the blue stain fungus, *Ophiostoma minus* (Hedgcock) H. & P. Sydow, which had been cultured on PDA medium at 25°C for 2 weeks. After incubation at 25°C under constant darkness for 4 weeks, a 1 ml suspension including 3000 nematodes of *B. mucronatus* alone (Bm treatment), *B. xylophilus* alone (Bx treatment), or *B. mucronatus* and *B. xylophilus* combined (Bm+Bx treatment) was poured into the hole of each bolt. Immediately after nematode inoculation, post-diapause *M. alternatus* larvae were placed into the holes singly and the openings of holes were plugged with rounded aluminium foil. The number of beetle larvae used in Bm, Bx, and Bm+Bx treatments was 25, 25, and 26, respectively. The bolts were kept at 25°C in the dark and checked daily for adult beetle emergence. Soon after emergence, adults were sexed and their body weight was determined. Adults were then crushed individually using a mortar and pestle, and the nematodes (the fourth-stage dispersal juveniles; a specialised developmental stage to travel between host trees) carried by them at emergence (initial nematode load) were extracted with the Baermann funnel technique at 25°C for 2-3 days.

To determine nematode density and the developmental stages, the bolts were chopped into four with a hatchet and the wood wall *ca* 0.5 cm thick was collected from the artificial pupal chamber just after the emergence of adults. Nematodes were extracted from chipped wall wood by using the Baermann funnel technique at 25°C for 2-3 days.

Of the nematodes extracted from the wood wall of artificial pupal chambers, about 50 were selected randomly and classified as third-stage dispersal juveniles (J3), fourth-stage dispersal juveniles (J4), and others including propagative juveniles and adults. The J3 were identified by the dark body coloration due to the presence of lipids (Mamiya, 1984). The J4 were determined by domed head and dark body. The number of J3 and J4 remaining in the artificial pupal chamber after beetle emergence was estimated as the product of the number of nematodes in the wood wall and the ratio of J3 and J4 in the sample. Consequently, the total number of dispersal juveniles for each pupal chamber was calculated as summation of initial nematode load and total number of J3 and J4 remaining in the pupal chamber. The proportion of successful moult to J4 was defined as the ratio of the total number of J4 to the total number of dispersal juveniles for each pupal chamber.

One-way analysis of variance (ANOVA) was conducted to test the differences in initial nematode load, total number of J3 and J4 remaining in the pupal chamber, total number of dispersal juveniles formed, and the proportion of successful moults to J4 among three different treatments. Following one-way ANOVA, the Tukey-Kramer multiple comparison test was used to compare means pairwise. Before analysis, the initial nematode load, the number of J3 and J4 remaining in the pupal chamber, and the total number of dispersal juveniles were logarithmically transformed to ensure their normality. The proportion of successful moult to J4 was arcsine-transformed for the same reason.

Results

A COMPARISON OF BOARDING ABILITY BETWEEN THE TWO NEMATODE SPECIES

Nineteen, 16, and 19 adults emerged from Bm, Bx, and Bm+Bx treatments, respectively. Mean initial nematode load was 316 (sd = 560), 23 669 (sd = 19 281), and 6982 (sd = 9048), for Bm, Bx, and Bm+Bx treatments, respectively (Table 1). Mean initial nematode load was significantly greater in Bx and Bm+Bx treatments than in Bm treatment (one-way ANOVA, $P < 0.05$, Tukey-Kramer multiple comparison test, $P < 0.05$).

Mean number of J3 and J4 that remained in pupal chambers just after beetle emergence was 26 165 (sd = 14 617), 10 039 (sd = 7412), and 18 239 (sd = 9344), for Bm, Bx, and Bm+Bx treatments, respectively (Table 1). It was significantly greater in Bm and Bm+Bx treatments than in Bx treatment (one-way ANOVA, $P < 0.001$, Tukey-Kramer multiple comparison test, $P < 0.05$).

The total number of dispersal juveniles was estimated as the sum of the initial nematode load and the number of J3 and J4 remaining in pupal chamber for each adult beetle. Mean of the total number of dispersal juveniles was 26 481 (sd = 14 797), 33 707 (sd = 22 839), and 25 222 (sd = 15 915), for Bm, Bx, and Bm+Bx treatments, respectively (Table 1). There was no significant difference in the mean of the total number of dispersal juveniles among the three different treatments.

There was a significant difference in the proportion of successful moults to J4 from J3 among the three treatments (one-way ANOVA, $P < 0.001$) (Table 1). Mean proportion of moulting to J4 was greatest in treatment Bx followed by Bm+Bx and Bm treatments in that order (Tukey-Kramer multiple comparison test, $P < 0.05$).

Table 1. *Interspecific competition of* Bursaphelenchus xylophilus *and* B. mucronatus *at the time of boarding* Monochamus alternatus *adults.*

Parameter		Treatment[1]		
		Bm	Bx	Bm+Bx
Beetles emerged (n)		19	16	19
Initial nematode load[2,3]	n	316 a	23 669 b	6982 b
	± SD	560	19 281	9048
Dispersal juveniles remaining in artificial pupal chamber[2,3]	n	26 165 a	10 039 b	18 239 a
	± SD	14 617	7412	9344
Total dispersal juveniles[2,3]	n	26 481 a	33 707 a	25 222 a
	± SD	14 797	22 839	15 915
Probability of moulting from J3 to J4[2,3,4]	*P*	0.022 a	0.613 b	0.222 c
	± SD	0.029	0.218	0.182

[1] Bm, Bx, and Bm+Bx represent *Pinus densiflora* bolts inoculated with *Bursaphelenchus mucronatus* alone, *B. xylophilus* alone, and *B. mucronatus* and *B. xylophilus* simultaneously, respectively.
[2] Mean ± SD.
[3] Means followed by the same letter are not significantly different at the 5% level (one-way ANOVA, Tukey-Kramer multiple comparison test).
[4] J3 and J4 indicate the third- and fourth-stage dispersal juveniles, respectively.

EVALUATION OF INTERSPECIFIC COMPETITION BETWEEN TWO NEMATODE SPECIES IN THE PROCESS OF BOARDING ON SAME VECTOR

Assuming a constant probability of moulting from one J3 to J4, an inhibitory effect of one nematode species on moulting of another species from J3 to J4 is expressed as follows:

$$P_x(1 - h_m) + P_m(1 - h_x) = P_{m+x},$$

where P_x, P_m, and P_{m+x} represent the probabilities of moulting from J3 to J4 in each treatment, respectively. h_m and h_x represent inhibitory coefficients by *B. mucronatus* and *B. xylophilus*, respectively. As our experiment gave $P_x = 0.613$, $P_m = 0.022$ and $P_{m+x} = 0.222$ (Table 1),

$$0.613(1 - h_m) + 0.022(1 - h_x) = 0.222.$$

Without a positive effect, the h_x value ranges from 0 to 1. Thus, the smallest h_m value was estimated to be 0.638 ($h_x = 1$) and the largest h_m

value was 0.674 ($h_x = 0$). The degree of inhibition by *B. mucronatus* on *B. xylophilus* moulting from J3 to J4, h_m, is estimated to be 0.6-0.7.

Discussion

The fourth-stage dispersal juvenile (J4) carried by vectors at their emergence (initial nematode load) is crucial to epidemiology of pine wilt disease. This study showed that *M. alternatus* adults carried a significantly larger number of *B. xylophilus* juveniles than *B. mucronatus* whereas there was no significant difference in the total number of dispersal juveniles between two nematode species. The proportion of successful moult to J4 was significantly greater in *B. xylophilus* than in *B. mucronatus*. Thus it is concluded that the significant difference in the initial nematode load between two nematode species reflects the difference in probability of moulting to J4. Necibi and Linit (1998) reported that the moult of *B. xylophilus* from J3 to J4 appeared to be related to adult eclosion of *M. carolinensis* (Olivier). They hypothesised that a genus-specific substance(s) associated with *Monochamus* adult eclosion ensures the *Monochamus-B. xylophilus* association. Our results suggested that there was a significant difference in the nematode response to the substance(s) associated with *M. alternatus* eclosion between the two nematode species.

The number of *B. xylophilus* transmitted to a pine tree might determine the rate of population increase after invasion of *P. thunbergii* trees (Hashimoto & Sanui, 1974). Initial nematode load is one of the most contributive factors to the variation in the number of nematodes transmitted per unit time in *M. saltuarius* (Gebler)-*B. mucronatus* system (Jikumaru & Togashi, 2001). The difference in the initial nematode load affects the number of nematodes transmitted to pine trees and persistence of the nematode in pine forests. The different boarding ability of two nematode species on a vector might be related to the replacement of *B. mucronatus* by *B. xylophilus* through the differences in initial nematode load and number of nematode transmitted to new host in pine forest ecosystems in Japan.

Evaluation of interspecific competition between two nematode species at a process of boarding on the same vector suggests a severe interspecific competition between *B. mucronatus* and *B. xylophilus* during the process of boarding on a common vector. The mechanism of this inter-specific competition remains to be understood.

References

AIKAWA, T. & TOGASHI, K. (1997). A simple method for loading adult *Monochamus alternatus* (Coleoptera: Cerambycidae) with *Bursaphelenchus xylophilus* (Nematoda: Aphelenchoididae). *Applied Entomology and Zoology* 32, 341-346.

HASHIMOTO, H. & SANUI, T. (1974). [The influence of inoculation quantities of Bursaphelenchus *lignicolus* Mamiya & Kiyohara on the wilting disease development in *Pinus thunbergii* Parl.] *Transactions of the Annual Meeting of the Japanese Forestry Society* 85, 251-253.

JIKUMARU, S. & TOGASHI, K. (2001). Transmission of *Bursaphelenchus mucronatus* (Nematoda: Aphelenchoididae) through feeding wounds by *Monochamus saltuarius* (Coleoptera: Cerambycidae). *Nematology* 3, 325-334.

KISHI, Y. (1995). *The pine wood nematode and the Japanese pine sawyer.* Tokyo, Japan, Thomas Co. Ltd, 302 pp.

LINIT, M.J. (1988). Nematode-vector relationships in the pine wilt disease system. *Journal of Nematology* 20, 227-235.

MAMIYA, Y. (1984). The pine wood nematode. In: Nickle, W.R. (Ed.). *Plant and insect nematodes.* New York, NY, USA, Marcel Dekker, pp. 589-626.

MAMIYA, Y. (1988). History of pine wilt disease in Japan. *Journal of Nematology* 20, 219-226.

MAMIYA, Y. & ENDA, N. (1979). *Bursaphelenchus mucronatus* n. sp. (Nematoda: Aphelenchoididae) from pine wood and its biology and pathogenicity to pine trees. *Nematologica* 25, 353-361.

NECIBI, S. & LINIT, M.J. (1998). Effect of *Monochamus carolinensis* on *Bursaphelenchus xylophilus* dispersal stage formation. *Journal of Nematology* 30, 246-254.

RUTHERFORD, T.A. & WEBSTER, J.M. (1987). Distribution of pine wilt disease with respect to temperature in North America, Japan, and Europe. *Canadian Journal of Forest Research* 17, 1050-1059.

Nematology Monographs & Perspectives, 2003, Vol. 1, 247-260

Pine wood nematode movement on, and migration from, the body surface of the Japanese pine sawyer during feeding on pine twigs after emergence

Akiomi YAMANE, Minamiko MIYOSHI, Yuichiro HATTA, Tomoko NAKAMURA, Ryûtarô IWATA, Toshihiko ISHIKAWA, Hirotaka KOBAYASHI and Yasuharu MAMIYA

College of Bioresource Sciences, Nihon University, Kameino, Fujisawa, Kanagawa, 252-8510 Japan

Summary – Movement of dauer larva of the pine wood nematode (PWN), *Bursaphelenchus xylophilus*, on the surface of the adult body of the Japanese pine sawyer (JPS), *Monochamus alternatus*, and its migration from the sawyer to pine twigs were observed. Observation of PWN was made regularly every day during the daytime, and sometimes at night, by counting numbers of PWN at specific parts of the insect body, mouth parts and surroundings, fore-, mid- and hind-legs, pair of the first abdominal spiracles, apical portion of abdomen, ventral surface of the abdomen and dorsal surface of the abdomen, using Keyences Digital High Definition Microscope. PWN was most abundant at the apical portion of abdomen especially when the twig surface was wet. During the day, PWN moved less when the temperature went up. On a cool day after a hot day more nematodes were moving. More PWN migrated from insects to wet than to dry twigs. Low temperature or temperature reduction activates PWN movement. This shows that the PWN dauer larvae on the sawyer body prefer wet conditions and a rather low temperature during migration from the sawyer.

With regard to the symbiosis between the pine wood nematode (PWN), *Bursaphelenchus xylophilus* (Nematoda: Aphelenchoididae) and the Japanese pine sawyer (JPS) *Monochamus alternatus* (Coleoptera: Cerambycidae) (Morimoto & Iwasaki, 1972; Yamane, 1981; Kobayashi *et al.*, 1984; Kishi, 1995) it is generally recognised that the numbers of PWN dauer juveniles migrating daily from JPS adults after emergence are closely related to the age of the JPS adult. In addition to age, chemical, physical and biological factors also play an important role in migration of the nematode, *i.e.*, temperature, humidity, oleoresin, *etc*. However, most of the process and factors of PWN movement on JPS adults after emergence and migration from JPS to pine twigs during feeding still remain to be solved.

We must begin by defining terms for nematode behaviour. Migration of nematode means nematode movement from the pupal cell wall of infested pine sapwood to a JPS callow adult, or from a JPS adult after emergence to a healthy standing pine during feeding through a feeding wound. On the other hand, movement of nematodes means movement of PWN on the JPS body surface.

Migration and movement of nematodes are one of the important subjects connected with the symbiotic relationship between JPS and PWN, and also essential for practical pest management of JPS with PWN. Among the important environmental factors affecting migration and movement of PWN, effects of temperature and water on the surface of pine twigs on which JPS adult feed, as well as the age of JPS, were studied experimentally.

Materials and methods

Japanese pine sawyer adults were collected from a field emergence cage every morning during the emergence season. They were kept individually in small containers with pine twigs as food. Every JPS had its own number made up of date of emergence/cage/sex/serial number of the cage.

Experiment 1: migration of nematodes on wet and dry twigs

To keep the pine twig always wet, we put wet cotton on the top portion of the twig on which the JPS adult was feeding. To keep the twig dry, it was supplied without wet cotton on it. Newly emerged JPS adults were collected from emergence cages and put individually in containers with wet or dry twigs on the Baermann apparatus. Nematodes which migrated to both the feeding twig and the bottom of the container were extracted daily with the routine Baermann apparatus.

Experiment 2: movement of nematodes on the JPS body surface

Direct observations of movement of nematodes on the adult body surface were made by counting numbers of nematodes on eight specific body parts: mouth parts and surroundings; fore-, mid- and hind-legs; pair of the first abdominal spiracles; apical portion of abdomen; ventral surface of abdomen; dorsal surface of abdomen; pronotum; and hind-wings of individual JPS, using a Digital High Definition Microscope Keyence. Observations were made three times a day for about 4 weeks. The times and durations were 10:00-11:30,13:00-14:30 and 16:00-17:30. For continuous daily observation, sample adults were fixed on a cork cap alive by using insect pins during observation.

Experiment 3: direct observation of nematodes on the JPS adult

On 19-31 July 2001, numbers of PWN on 11 adult JPS were observed daily. Most JPS used were about 14 days old, ranging from 3 to 14 days. Direct observations were made as ambient temperature decreased from 30°C to 25, 27.5 and 20°C for days 1-4, 5-6 and 7-10. Each of the first daily observations was made at 30°C, and the next observation made when the temperature decreased. When observations were completed, the room temperature was raised to 30°C. Both number of nematodes and nematode mass were counted at the same time, classifying masses into small-, medium-, and large-sized, containing about 10, 150, and 500 nematodes, respectively. Occurrence of mass formation was expressed as means of frequency per day.

Experiment 4: Comparison of nematode migration at 30 and 20°C

Numbers of nematodes migrating in a daily temperature regime of 30°C for 16 h and 20°C for 8 h were assessed by extracting nematodes from feeding twigs by Baermann apparatus for 5 days.

Results

More nematodes migrated in wet than in dry conditions, and the peaks of migrating PWN from individual JPS occurred earlier in wet than

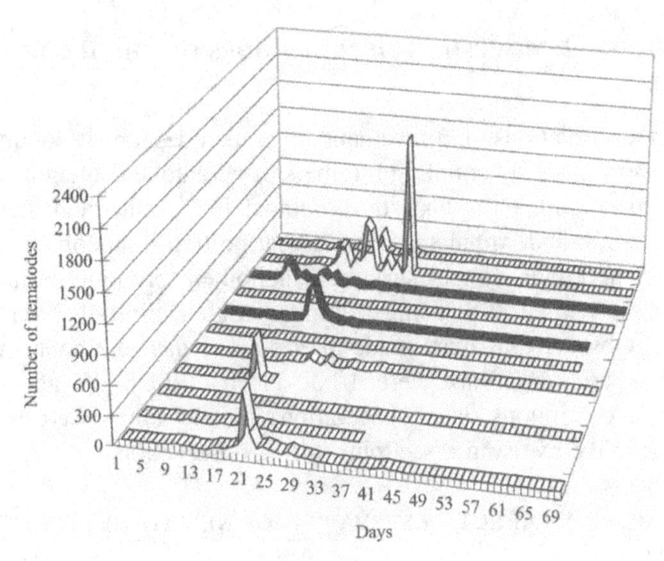

Fig. 1. *Numbers of* Bursaphelenchus xylophilus *migrated from 11 Japanese pine sawyers per day after sawyer emergence in dry conditions.*

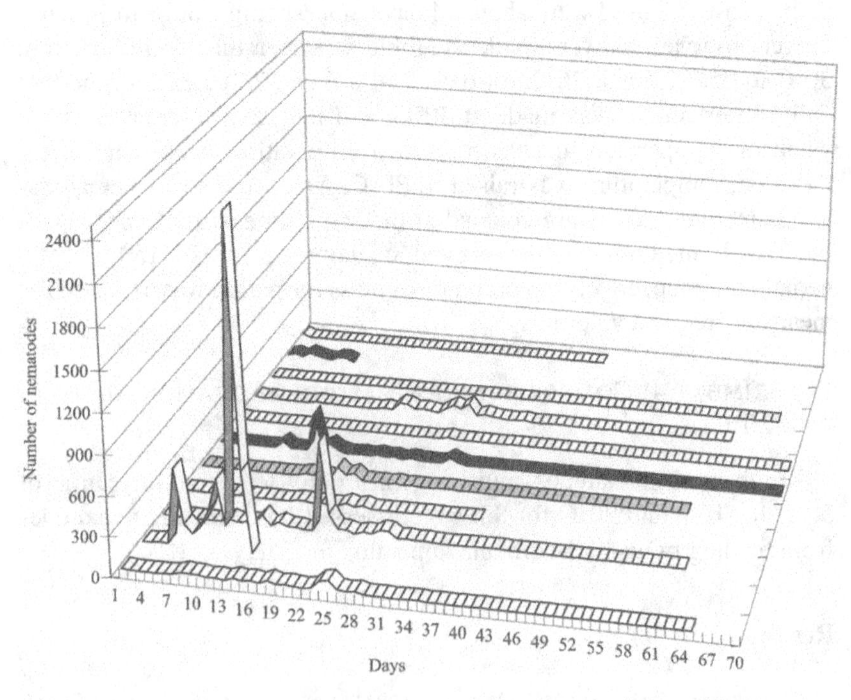

Fig. 2. *Numbers of* Bursaphelenchus xylophilus *migrated from 11 Japanese pine sawyers per day after sawyer emergence in wet conditions.*

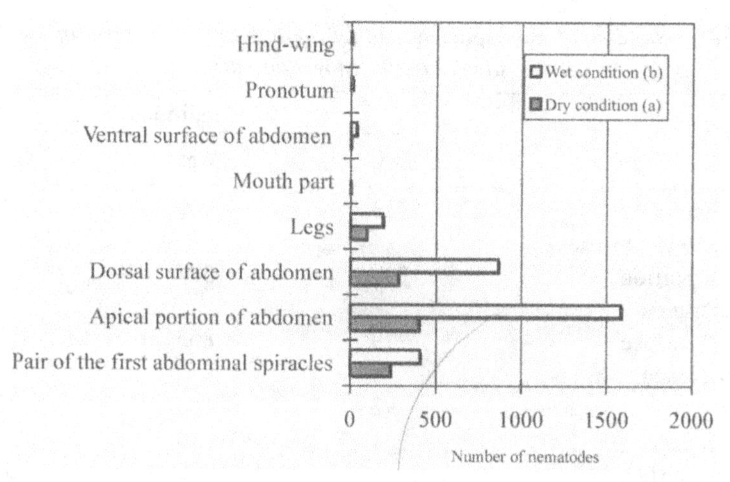

Fig. 3. *Total number of* Bursaphelenchus xylophilus *on specific parts of the body surface of adult Japanese pine sawyers in wet (□) and dry (■) conditions.*

in dry conditions (Figs 1, 2). However, it is difficult to say that age dependency was clear, because of great individual variation.

Daily observations of PWN numbers at certain parts of JPS adult body showed that most of the PWN on JPS body were detected at the apical portion of the abdomen (Table 1; Fig. 3). PWN movement seemed to be mainly from the pair of the first abdominal spiracles towards the apical portion of the abdomen, as supposed in previous papers (Hosoda *et al.*, 1978).

Numbers of PWN on JPS at three observation times during daytime were not very different, but the numbers of PWN moving in wet conditions were greater than in dry conditions (Figs 4, 5; Table 2).

According to the 24 h observation at 3 h intervals, a drastic decrease in the number of PWN was observed between 19:00 and 22:00, both in wet and dry conditions. During the period, the temperature went up by about 2-3°C (Fig. 6). Figs 7 and 8 show daily fluctuations of the numbers of PWN moving which seemed to be connected to changes of mean daily temperatures. On the days after mean temperature went down from about 30 to 25°C, the numbers of moving PWN clearly increased.

When temperatures decreased, the numbers of moving PWN which per day were 4.1, 0.9 and 178.3, at 25, 27.5 and 20°C, respectively, and PWN masses were seen in 36.0, 27.7 and 51.2% of all cases (Table 3). The lower the temperature, the more frequently masses were formed.

Table 1. *Number of* Bursaphelenchus xylophilus *on eight parts of the body surface of Japanese pine sawyer adults in wet and dry conditions.*

Body part	Conditions		
	Dry	Wet	Wet/dry
First pair of abdominal spiracles	233	406	1.7
Apical portion of abdomen	399	1584	4.0
Dorsal surface of abdomen	279	865	3.1
Legs	92	189	2.1
Mouth part	1	0	–
Ventral surface of abdomen	1	33	33.0
Pronotum	–	8	–
Hindwing	–	2	–
Total	1005	3087	43.9

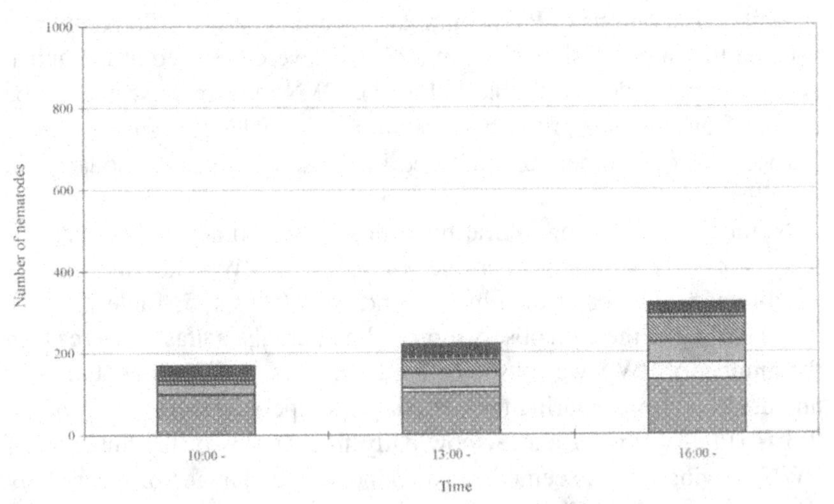

Fig. 4. *Numbers of* Bursaphelenchus xylophilus *on the bodies of ten Japanese pine sawyer adults at three times of day in dry conditions.*

As seen in Table 4, the number of PWN moving per h at 20°C was about 3.4 times more than at 30°C.

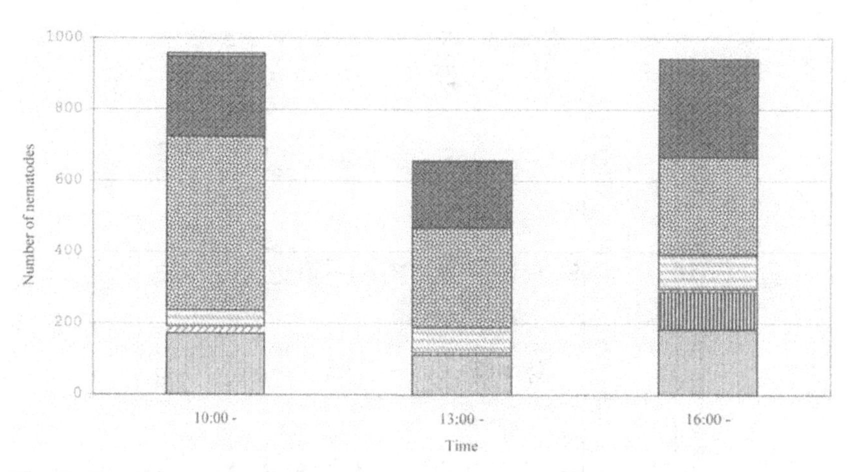

Fig. 5. *Numbers of* Bursaphelenchus xylophilus *on the bodies of ten Japanese pine sawyer adults at three times of day in wet conditions.*

Discussion

PWN transmission to new host pine trees by JPS is realised through the following eight steps: *i*) aggregation of PWN around pupal cell wall of JPS in infested pine; *ii*) migration of PWN from cell wall in the sapwood of infested pine onto the callow adult of JPS in the pupal cell; *iii*) movement of PWN on the surface of callow adult of JPS body to the spiracles, mainly the first abdominal spiracle, and metathoracic spiracle; *iv*) movement of PWN into the tracheae; *v*) exit of PWN from the spiracles; *vi*) movement of PWN on the surface of adult body; *vii*) aggregation of PWN at the tip of abdomen; and *viii*) migration of PWN from JPS to living pine twigs. In this report we try to clarify the processes and factors involved in steps *v* through *viii*.

With regard to the processes of nematode migration, it has been shown that 25% of the nematodes held by the emerged adults migrated from the sawyer within 10 days, 87% within 20 days, and 94% within 30 days after the emergence when the temperature was 25°C (Morimoto & Iwasaki, 1971; Enda, 1972). Adults 25 or more days after emergence did not carry nematodes, and adults which had emerged in July carried fewer nematodes less frequently than those which had emerged in June (Taketani *et al.*, 1974). The number of nematodes retained by a JPS decreases after its emergence; usually migration is uncommon within a week of JPS emergence and reaches a maximum after 2-3 weeks, then

Table 2. *Numbers of* Bursaphelenchus xylophilus *moving on the body surface of Japanese pine sawyer adults over 24 h.*

Conditions		Date	Time (h)								Total
			10:00	13:00	16:00	19:00	22:00	01:00	04:00	07:00	
Dry		21/7	7	11	11	35	8	3	0	1	–
		31/7	16	13	66	51	28	32	6	5	–
		7/8	5	21	12	75	33	9	5	0	–
	Total		28	45	89	161	69	44	11	6	–
	Average		9.3	15.0	29.7	53.7	23.0	14.7	3.7	2.0	151.0
Wet		21/7	37	31	21	43	5	1	4	0	–
		31/7	175	161	133	158	62	27	29	20	–
		7/8	129	79	145	197	19	10	11	4	–
	Total		341	271	299	398	86	38	44	24	–
	Average		113.7	90.3	99.7	132.7	28.7	12.7	14.7	8.0	500.3
Wet/dry			12.2	6.0	3.4	2.5	1.2	0.9	4.0	4.0	3.3
Average temperatures (°C)			25.0	25.0	24.7	24.3	26.3	26.7	26.7	26.7	–

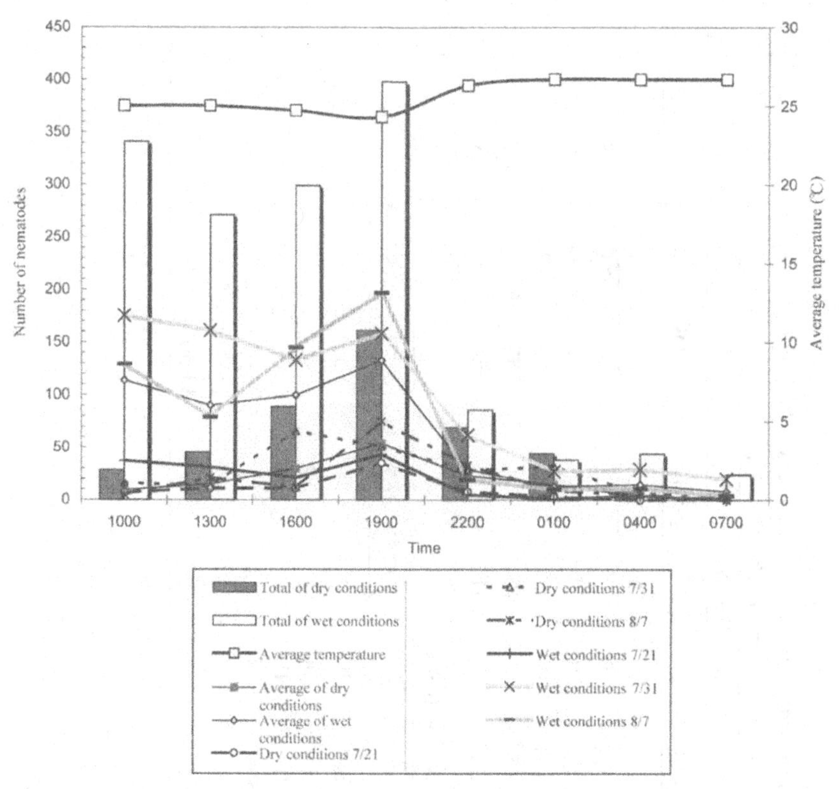

Fig. 6. *Numbers of* Bursaphelenchus xylophilus *moving on the body surface of Japanese pine sawyer adults during 24 h and temperature.*

gradually declines and terminates by weeks 4 or 5 (Enda, 1972; Mineo & Kontani, 1975; Nakane, 1976; Hosoda & Kobayashi, 1978).

On the other hand, Kishi (1978, 1980) showed that the most active migration occurs within the first 5 days after JPS emergence, but at the same time he showed the most active movement took place from 6-15 days after emergence.

With regard to movement of PWN on JPS adult body, Enda (1973) showed that PWN moved out from the spiracles and moved toward the apical portion of abdomen.

Volatile materials, β-myrcene in particular, were reported to play a large role in transporting PWN from JPS into trees (Ishikawa *et al.*, 1984; Ishikawa *et al.*, 1986). Kibe *et al.* (2001) also showed that some of the monoterpenes in the pine oleoresin activate movement of PWN.

Table 3. *Average numbers of nematodes moving on individual Japanese pine sawyers (JPS) decreased after temperature each day from 30 to 27.5°C for 2 days, 25°C for 4 days and 20°C for 4 days (July 19-31).*

Temper-ature (°C)	No. of JPS											
	7/5C♂-175		7/5C♂-178		7/5C♂-182		7/5C♀-101		7/5C♀-103		7/5C♀-113	
	a	b	a	b	a	b	a	b	a	b	a	b
25	8.5	75	15.0	100	6.0	25	0	0	0.3	0	3.3	25
27.5	1.0	50	2.0	50	6.0	50	0	0	0	0	1.0	50
20	636.0	80	460.7	57	799.8	100	1.9	29	0.9	29	3.8	50
c	750		3450		17 850		299		257		38	
d	3969		6739		21 085		312		264		65	

	No. of JPS									
	7/6C♀-127		7/7C♂-226		7/9A♂-48		7/9C♀-167		7/21A♂-187	
	a	b	a	b	a	b	a	b	a	b
25	0	0	5	25	5	100	2	50	0	0
27.5	0	0	0	0	0	0	0	50	0	0
20	1.1	14	7.8	75	5.9	29	0.4	14	43.9	86
c	211		98		–		0		11 000	
d	219		144		56		9		11 307	

a: Number of nematodes on JPS per day.
b: Occurrence of nematode masses per day (%).
c: Number of nematodes remaining on JPS.
d: Total number of nematodes on JPS.

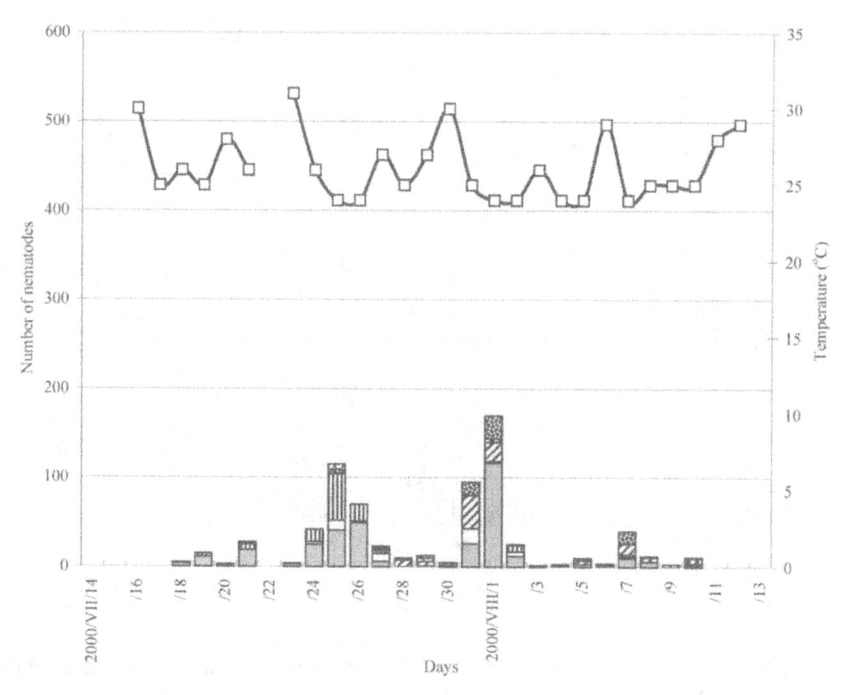

Fig. 7. *Relationship between temperature change and numbers of* Bursaphelenchus xylophilus *moving on the body surface of ten Japanese pine sawyers in dry conditions and temperature at 13:00 h (□).*

Togashi (1985) reported that the peak of invading PWN changed greatly according to the number of PWN carried by sawyers. The more nematodes that were carried, the earlier the invasion peak took place: 15-25 days after emergence with more than 10 000 nematodes, 30-35 days after with 1000-10 000, and 20-25 days after with 100-1000 nematodes.

The present reports did not show detailed environmental conditions, but showed that not only age but environmental factors and age together may play an important role in migration of PWN.

We may conclude that movement and direction of PWN on the JPS body were clearly observed as follows: PWN in the spiracles moved out, mainly to the apical portion of abdomen of JPS after emergence. There was more movement and migration of PWN in wet than in dry conditions; most of the PWN migrate from the apical portion of the abdomen of JPS, and nematode mass formation occurred more frequently at low temperatures. Low or falling temperatures accelerated PWN movement and migration. Considering this, PWN migration from

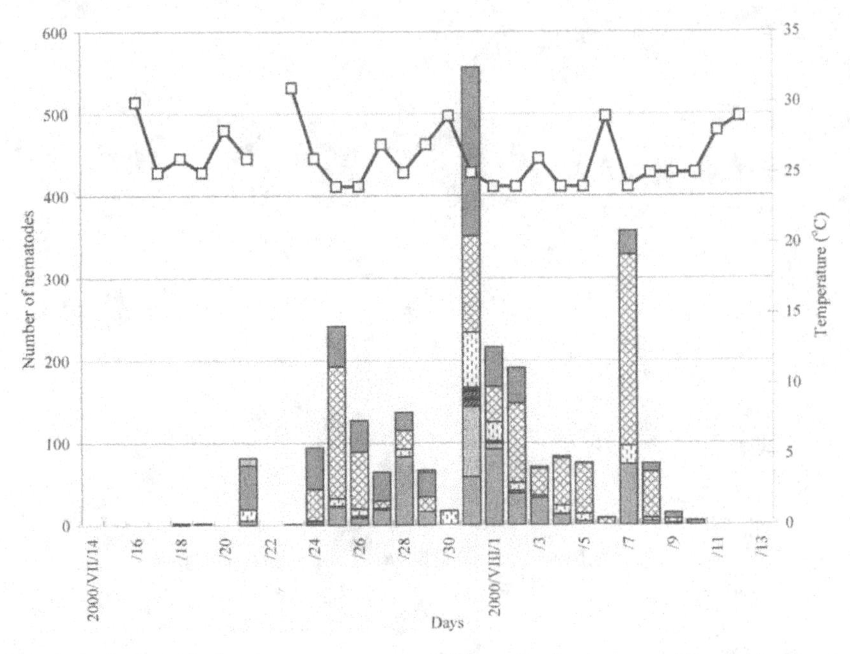

Fig. 8. *Relationship between temperature change and numbers of* Bursaphelenchus xylophilus *moving on the body surface of ten Japanese pine sawyers in wet conditions and temperature at 13:00 h (□).*

Table 4. *Numbers of* Bursaphelenchus xylophilus *migrating from Japanese pine sawyer adults and comparison of rates of movement at 30 and 20°C.*

Adult	Temperature (°C)				Hourly rate 30/20
	30		20		
	16 h	/h	8 h	/h	
7/26A♀-175	215	13.43	517	64.63	4.81
-177	171	10.69	685	85.63	8.01
-178	199	12.44	204	25.5	2.05
-182	668	41.75	694	86.75	2.08
-183	1931	120.69	1228	153.5	1.27
-184	159	9.94	112	14	1.41
-275	1461	91.31	4048	506	5.54
7/28A♀-205	140	8.75	236	29.5	3.37
7/29A♀-209	481	30.06	1398	174.75	5.81
7/29A♂-249	1180	73.75	1962	245.25	3.33
Total	6605	412.81	11 084	1385.51	3.36

JPS after emergence depended not only on JPS adult age, but also on various physical, chemical and environmental factors, particularly temperature and water.

References

ENDA, N. (1972). Drop-off procedure of the pine wood nematodes from the pine sawyer. *Transactions of 24th Annual Meeting of Kanto Branch, Japanese Forestry Society*, p. 32.

ENDA, N. (1973). Nematode numbers on various parts of the JPS adult body. *Transactions of 25th Annual Meeting of Kanto Branch, Japanese Forestry Society*, p. 16.

HOSODA, R. & KOBAYASHI, K. (1978). Drop-off procedures of the pine wood nematodes from the pine sawyer (2). *Transactions of 29th Annual Meeting of Kanto Branch, Japanese Forestry Society*, pp. 131-132.

ISHIKAWA, M., HINODE, Y., SHUTO, Y., MATSUMORI, K. & WATANABE, H. (1984). Bio-organic chemical studies on pine withering caused by pine wood nematode *Bursaphelenchus lignicolus. Abstracts of symposium on biometric chemistry at Kyushu University*, pp. 9-12.

ISHIKAWA, M., SHUTO, Y. & WATANABE, H. (1986). β-myrcene, a potent attractant component of pine wood for the pine wood nematode, *Bursaphelenchus xylophilus. Agricultural Biology and Chemistry* 50, 1863-1866.

KIBE, N., YAMANE, A., IWATA, R. & ISHIKAWA, T. (2001). Effects of monoterpene on movement of PWN from JPS during feeding on pine twig. *Abstracts of 112th Annual Meeting of the Japanese Forestry Society*, p. 313.

KISHI, Y. (1978). Invasion of pine trees by *Bursaphelenchus lignicolus* M. & K. (Nematoda; Aphelenchoidae) from *Monochamus alternatus* Hope (Coleoptera; Cerambycidae). *Journal of Japanese Forestry Society* 60, 179-182.

KISHI, Y. (1980). Mortality of pine trees caused by *Bursaphelenchus lignicolus* M. & K. (Nematoda; Aphelenchoidae) in Ibaraki Prefecture and its control. *Bulletin of Ibaraki Forestry Experimental Station* 11, 1-83.

KISHI, T. (1995). *The pine wood nematode and the Japanese pine sawyer.* Tokyo, Japan, Thomas Co Ltd, 302 pp.

KOBAYASHI, F., YAMANE, A. & IKEDA, T. (1984). The Japanese pine sawyer beetle as the vector of pine wilt disease. *Annual Review of Entomology* 29, 115-135.

MINEO, K. & KONTANI, S. (1975). On the seasons of nematode movement from the pine sawyer into pine trees. *Transactions of the 86th Annual Meeting of the Japanese Forestry Society*, pp. 307-308.

MORIMOTO, K. & IWASAKI, A. (1971). Transmission of the pine wood nematode by the pine sawyer. *Transactions of the 25th Annual Meeting of the Kyushu Branch, Japanese Forestry Society*, pp. 165-166.

MORIMOTO, K. & IWASAKI, A. (1972). Role of *Monochamus alternatus* (Coleoptera; Cerambycidae) as a vector of *Bursaphelenchus lignicolus* (Nematoda: Aphelenchoidae). *Journal of Japanese Forestry Society* 54, 177-183.

NAKANE, I. (1976). Drop-off of the dauer larvae of the pine wood nematode from the pine sawyer. *Transactions of the 27th Annual Meeting of the Kansai Branch, Japanese Forestry Society*, pp. 252-254.

TAKETANI, A., GISIDAM, R. & OKUDA, M. (1974). Feeding amounts of the pine sawyer adults. *Transactions of the 25th Annual Meeting of the Japanese Forestry Society*, pp. 302-305.

TOGASHI, K. (1985). Transmission curve of *Bursaphelenchus xylophilus* (Nematoda; Aphelenchoidae) from its vector, *Monochamus alternatus* (Coleoptera: Cerambycidae) to pine trees with reference to population performance. *Applied Entomology and Zoology* 20, 246-251.

YAMANE, A. (1981). The Japanese pine sawyer, *Monochamus alternatus* Hope (Coleoptera: Cerambycidae): bionomics and control. *Review of Plant Protection Research* 14, 1-25.

Nematology Monographs & Perspectives, 2003, Vol. 1, 261-266

Latent infection of *Bursaphelenchus xylophilus* and a new means of transmission by *Monochamus alternatus*

Bao Jun YANG [1], Lai Fa WANG [1],
Fu Yuan XU [2] and Pei ZHANG [2]

[1] *The Chinese Academy of Forestry, Beijing, 100091 China*
[2] *Forestry Academy of Jiangsu, Nanjing, Jiangsu, 211153 China*

Summary – Felling pine trees affected by pine wilt disease is used to control disease spread but apparently unaffected trees may carry latent infections of pine wood nematodes (PWN). In experiments inoculating seven species of 15-20-year-old pine trees with PWN, latent infection was common, particularly in the more resistant species. Adult pine beetles emerging from larvae that had been reared in pine branches with latent infections subsequently transmitted PWN through feeding on healthy branches. This indicates the importance of early diagnosis of infection to eliminate pine wilt.

In pine wood nematode (PWN) diseased areas, felling infested trees is widely used as a control method. Felling dead pines can reduce the beetle population and eliminate PWN. However, despite felling wilted pines, other pines have died in following years, suggesting latent infections by this nematode. This research was to determine if PWN latent infections occurred and in what conditions. The experiment tested to see if beetles, reaching maturity by feeding on pine trees containing PWN but not showing symptoms of wilt, could transmit the nematodes to healthy pines.

Material and methods

LATENT INFECTION TEST

The pines used were seven species of 15-20-year-old *Pinus*, Japanese black pine (*Pinus thunbergii*), Japanese red pine (*P. densiflora*), masson pine (*P. massoniana*), pitch pine (*P. rigida*), slash pine (*P. elliottii*), loblolly pine (*P. taeda*) and shortleaf pine (*P. echinata*).

PWN were collected from Japanese black pine in Nanjing and cultured on *Pestalotia* sp. Inoculation was made by injecting a water suspension of PWN into holes made in the stem of pines at a rate of 5000 nematodes per pine. The following year, samples were taken from inoculated pines at 10 cm above the inoculation sites. Nematodes were isolated and identified from the samples.

BEETLE TRANSMISSION TEST

Nematode-free branches were cut from healthy masson pine and Japanese black pine. The branches were placed in flasks full of water, two branches in one flask, at room temperature (*ca* 25-30°C). Diseased branches, taken from the same pine species, were produced by inoculating them with PWN. After 20 days, we isolated PWN from the phloem and xylem of infected and healthy branches to confirm that there were no nematodes in nematode-free branches and that there were nematodes in diseased branches. Five branches were taken randomly for each.

Beetle larvae were collected from Japanese black pines in Nanjing. The larvae were washed in disinfected water and put in artificial holes on a healthy Japanese black pine trunk taken from a PWN-free area (Wu County). After the beetles emerged, eight beetles sampled at random were dissected and isolated to guarantee PWN-free beetles.

The PWN-free beetles and diseased branches of Japanese black pine and masson pine were put into cages in which beetles were fed until maturity. After 2, 4, 6, 8 and 10 days, the diseased branches were taken out, healthy branches were put in, and beetles allowed to continue feeding for 10 days. Eleven beetles feeding on diseased branches for different times were sampled at random and dissected to isolate PWN. Ten days later the healthy branches were put into room temperature for 15 days, then nematodes isolated from these branches.

Results

Latent infection of PWN was common and varied as shown in Table 1. In a suitable environment, susceptible pines, such as Japanese black pine and red pine, died soon after inoculation. When the inoculum dose was

Table 1. *Survival of trees and numbers of pine wood nematodes (PWN) in seven* Pinus *species after inoculation with PWN at different doses at two dates.*

Pine species	Time of inoculation (date in 1997)	Time of observation (date in 1998)	Inoculum (PWN/ branch)	Test trees	Live trees	PWN infested trees	PWN (number/ g)
Japanese black pine	11/5	20/5	100	5	1	0	0
			500	5	0	0	0
			5000	5	0	0	0
	20/8		100	5	3	1	18
			500	5	2	1	3
			5000	5	0	0	0
Red pine	11/5	20/5	5000	5	0	0	0
	20/8		5000	4	2	0	0
Masson pine	11/5	20/5	100	10	10	2	155
			500	10	9	3	35
			5000	10	10	5	167
	20/8		100	10	8	0	0
			500	10	6	2	222
			5000	10	5	1	1413
Pitch pine	11/5	20/5	5000	5	5	0	0
	20/8		5000	5	4	1	108
Slash pine	11/5	20/5	5000	10	10	7	154
	20/8	20/5	5000	10	10	5	757
Loblolly pine	11/5	20/5	5000	10	10	2	1168
	20/8		5000	10	10	6	1541
Shortleaf pine	11/5	20/5	5000	5	5	0	0
	20/8		5000	5	5	0	0

very small, or the weather was getting colder, latent infection could occur and a few living pines contained fewer PWN. For resistant pines, such as loblolly pine and slash pine, latent infection was more common and there were more PWN in the samples. For moderately resistant pines, such as masson pine and pitch pine, there was a medium universal level of latent infection and of PWN numbers in the samples.

Table 2. *Pine beetles carrying pine wood nematode (PWN) after feeding on PWN diseased pine branches.*

Feeding time on diseased branched (days)	Number of beetles tested	Number of beetles carrying PWN	PWN number in each beetle
2	1	1	400
4	3	1	230
6	4	1	500
8	2	0	0
10	1	1	100

Table 3. *Transmission of pine wood nematode after initially nematode-free beetles fed on PWN infected branches of two pine species.*

Pine	Feeding time on diseased branches (days)	Number of beetles	Number of healthy branches	Number of branches with PWN	PWN (number/g)
Japanese black pine	2	4	2	0	0
	4	4	2	1	3.5
	6	4	2	1	7
	8	2	1	0	0
	10	3	2	1	11
Masson pine	2	2	1	0	0
	4	2	1	1	1.7
	6	2	–	–	–
	8	2	1	0	0
	10	2	–	–	–

After inoculation, shortleaf pine did not wilt and there were no nematodes in the samples. Shortleaf pine was therefore considered to be immune to PWN.

Eleven PWN-free beetles feeding on PWN infected pine branches for different times were sampled at random, dissected and PWN isolated. The results showed that about 40% of beetles carried PWN. There was no interrelation between the feeding time, the number of beetles carrying PWN and PWN quantity.

The beetles feeding on PWN infected pine branches for 2, 4, 6, 8 and 10 days, respectively, were put on nematode-free pine branches and continued feeding. Table 3 shows the results. About 15% of beetles transmitted PWN from diseased to healthy pine branches. There was no correlation between this transmission ability of beetles and feeding time on diseased branches.

Discussion

Felling pine infected by PWN is used widely to control pine wilt disease in PWN diseased areas. After felling wilted pines, however, more pines died the following next year. The results suggest that this phenomenon is related to latent infection of PWN. So, in a new diseased place, when wilted pines are felled, early diagnosis is needed to identify pines infected by PWN but not yet wilting. The purpose is to completely eliminate diseased pines.

In the experiment on beetle transmission of PWN, 200 beetle larvae were put in the artificial holes, but only 46 adult beetles emerged. This could be because the holes were not natural and were therefore unsuitable for beetle emergence. Also, the adult beetles were not very healthy as seven died during the experiment. The results showing that about 15% of beetles transmitted PWN from diseased pines to healthy pines by feeding indicate that early diagnosis is very important for elimination of diseased pines.

References

BERGDAHL, D.R. & HALIK, S. (1998). Inoculated *Pinus sylvestris* serve as long-term hosts for *Bursaphelenchus xylophilus*. In: Futai, K., Togashi, K. & Ikeda, T. (Eds). *Sustainability of pine forests in relation to pine wilt and decline. Proceedings of the symposium, Tokyo, Japan, 26-30 October 1998.* Kyoto, Japan, Shokado Shoten, pp. 73-78.

KIYOHARA, T. & TOKUSHIGE, Y. (1971). Inoculation experiments of a nematode, *Bursaphelenchus* sp., onto pine trees. *Journal of the Japanese Forestry Society* 53, 210-218.

LIU, W., YANG, B.J., XU, F.Y., GE, M.H., ZHANG, P. & WANG, C.F. (1998). [Study on early diagnosis for pine wilt disease caused by *Bursaphelenchus xylophilus*. I. Early diagnosis for pine wilt disease caused by *Bursaphelenchus xylophilus* parasitizing *Pinus massoniana* and *P. thunbergii*.] *Forest Research* 11, 455-460.

MAMIYA, Y. (1983). Pathology of the pine wilt disease caused by *Bursaphelenchus xylophilus*. *Annual Review of Phytopathology* 21, 201-220.

YANG, B.J., LIU, W., XU, F.Y., ZHANG, P. & QU, H.R. (1999). [Study on early diagnosis for pine wilt disease caused by *Bursaphelenchus xylophilus*. II. The effect of pine species, dose and nematode origin on the oleoresin exudation method.] *Forest Research* 12, 251-255.

Nematology Monographs & Perspectives, 2003, Vol. 1, 267

An observation of *Monochamus scutellatus* associated with conifer trees damaged or killed by wildland fire

Dale R. BERGDAHL

Department of Forestry, University of Vermont, Burlington, 05405 Vermont, USA

Summary – In eastern North America, the white-spotted sawyer beetle, *Monochamus scutellatus* (Say), is a wood-boring insect that is usually responsible for causing extensive damage to recently felled logs or standing injured or dead coniferous trees (especially, pine, spruce and fir). Adult beetles are most abundant from late May through July when they attack trees for the purpose of oviposition. This beetle is known to transmit the pine wood nematode (*Bursaphelenchus xylophilus* (Steiner & Buhrer) Nickle) during feeding and oviposition. Adult beetles are charcoal-black with a white spot at the base of their wings. This black colour resembles the charred bark or wood of fire-killed trees. During late June 1996, I was on a wildland fire crew assigned to the Graham Fire Complex located in a remote area of west central Ontario, Canada. The area burned included both black spruce (*Picea mariana* (Mill.) B.S.P.) and jack pine (*Pinus banksiana* Lamb.) fire types and extended over 30 000 acres. During and immediately following the active burn, very many *M. scutellatus* were observed in flight, and found breeding on and attacking fire-damaged or killed trees. However, beetles were not observed in areas where crown fires killed trees. Apparently, excessive heat generated during crown fires destroys the trees' attractiveness to *M. scutellatus*. In areas where either ground or surface fires predominate, we should expect an increase in populations of *M. scutellatus* and, if present, pine wood nematode populations would also be expected to increase.

Nematology Monographs & Perspectives, 2003, Vol. 1, 269-281

Successful control of pine wilt disease in Fukiage-hama seacoast pine forest in southwestern Japan

K. NAKAMURA [1] and N. YOSHIDA [2]

[1] *Kyushu Research Center, Forestry and Forest Products Research Institute, Kumamoto 860-0862, Japan*
[2] *Forestry and Forest Products Research Institute, Ibaraki 305-8687, Japan*

Summary – Control efforts against pine wilt disease (PWD) in Japan have not been fruitful on a nationwide scale; on the other hand, there are many successful examples locally. In Fukiage-hama seacoast pine forest, southwestern Japan, a PWD outbreak occurred in 1992, while the conventional control operations had been conducted. The damage area spread widely in 1994 and the main factors for this epidemic were considered to be *i)* the incomplete eradication of the trees infested by the insect vector, and *ii*) incompatibility of the timing of aerial spraying of insecticide and the insect emergence time. The improved control operation was performed from the autumn of 1994 and the outbreak subsided by 1997. Lessons from the failure and success of controlling PWD in Fukiage-hama would be useful for other areas to cope with this disease.

Pine wilt disease (PWD) has severely damaged pine forests in Japan (Mamiya, 1988; Kishi, 1995). Despite a great deal of control effort, spread of the disease within the country has not been prevented and the annual loss of pine trees is still as much as *ca* 1 000 000 m^3 per year (Fig. 1). We can state that pine wilt damage would be much more severe if the control efforts had not been made. However, it must be frankly stated that the eradication of PWD in Japan on a nationwide scale has been unsuccessful. On the other hand, we do have many successful examples of controlling PWD locally. Because the control operation against pest organisms is very empirical, we believe the Japanese experiences of failure and success in controlling PWD could offer lessons for countries where PWD is newly introduced.

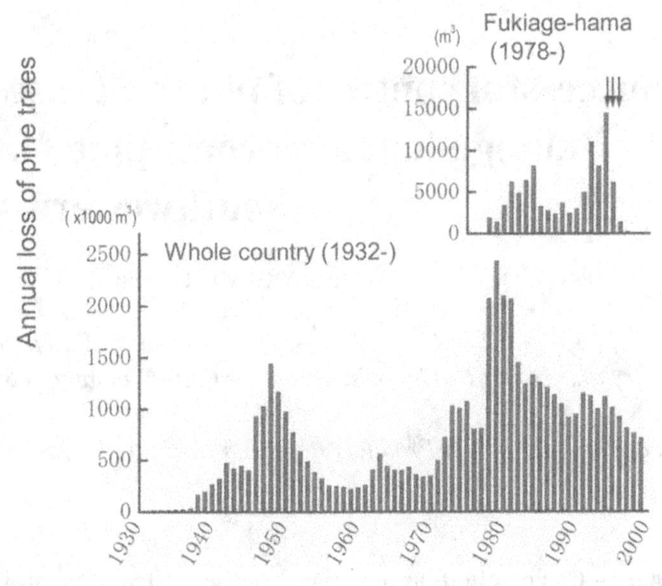

Fig. 1. *Damage level of pine trees in Japan and in Fukiage-hama alone. Data were available from 1932 and 1978 for the whole country and Fukiage-hama, respectively. Arrows in the upper graph indicate the implementation of the special control operations.*

In this paper, we first give a brief introduction of the control methods used against PWD adopted practically in Japan. Then, we report on an outbreak of PWD following an incomplete control operation and successful control of this epidemic with a great deal of effort in Fukiage-hama, a seacoast pine forest in southwestern Japan (Fig. 2).

Control methods against pine wilt disease in Japan

The pine wood nematode, *Bursaphelenchus xylophilus*, is the causal agent of PWD (Kiyohara & Tokushige, 1971), and is mainly vectored by the Japanese pine sawyer, *Monochamus alternatus*, in Japan (Mamiya & Enda, 1972; Morimoto & Iwasaki, 1972). Because of the difficulty of controlling the nematode itself, most of the control methods target the insect vector.

The flight season of *M. alternatus* is late spring through summer in most areas in Japan (Kobayashi *et al.*, 1984). The adults feed on the bark of pine twigs throughout their life and the females lay eggs in the inner bark of the weakened or dying pine tree. The hatched larva grows feeding

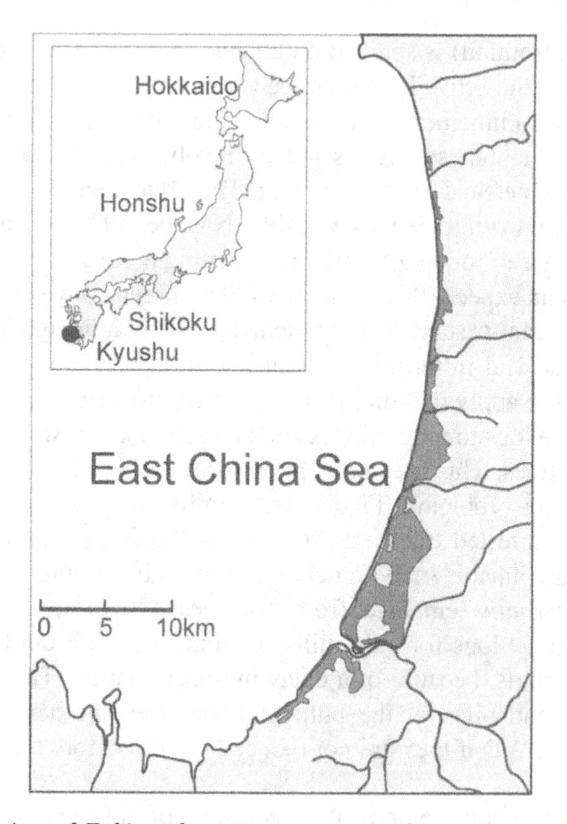

Fig. 2. *Location of Fukiage-hama seacoast pine forest. The dark grey area shows the range of Fukiage-hama National Forest, and there were small pine forests around this area.*

on the inner bark and sapwood of the pine tree, then bores a shallow tunnel for overwinter and pupation. The eclosed adult stays within the tree for about 1 week and emerges from the tree in May through July. Since immature *M. alternatus* stay inside the infested tree for a relatively long time, disposal of the trees to kill the immatures in it is a basic and major tactic for eradicating *M. alternatus*.

Incineration has traditionally been adopted in control operations against PWD. This simple method is highly effective, but requires much time and labour. Because of a labour shortage in rural areas, it becomes difficult to use this method in a broad situation in Japanese pine forests. The other problem for this method is that it entails the risk of fire.

At present, the most popular way of disposing of the infested trees is the application of infiltrative pesticide. An oil solution of insecticide (mostly fenitrothion) is sprayed on the bark of the cut pine logs and it is expected that the active ingredient penetrates under the bark and into the xylem killing immature *M. alternatus*. This method is adopted widely because it is labour saving, as it only involves cutting down the trees and applying pesticide with a sprayer. The effectiveness of this method is, however, known to be very variable (Nakane, 1977; Kobayashi *et al.*, 1981). The use of fumigant (carbam) is getting more and more popular in Japan. We can expect 100% mortality of the insects as well as nematodes within fumigant-treated logs (Tsuchiya, 1985), although this method costs a great deal in terms of labour and money. The cut logs need to be gathered to apply the fumigant and sealed properly.

Taking the logs to pulp mills is another very popular way of disposing of infested trees. This method is good from the viewpoint of recycling, but has some problems. Firstly, pulp mills only take large logs, so branches of infested trees are often left in the forest. Immature *M. alternatus* may inhabit such branches and grow into adults. We observed that *M. alternatus* emerged from branches of *ca* 2 cm diam. in the field. Secondly, logs for pulp mills are sometimes left until after insect emergence starts, because of a delay in transportation. Thirdly, the logs transferred and piled at the pulp mill may become a new source of infection of PWD if they are not reduced to pulp before the insect flight season.

There are other methods for disposing of infested trees, such as chipping and burying under the ground.

The aerial spraying of insecticide (mostly fenitrothion) is adopted for the purpose of minimising the transmission of *B. xylophilus* by adult *M. alternatus* feeding on twigs of live pine trees. Therefore, the timing of application is adjusted to hit the early stage of the insect flight season. The insecticide on pine twigs remains effective for a few weeks and the adults eating the twigs will be killed. If properly applied, this method is very effective in preventing the spread of PWD. There are, however, some critical limitations for this method. To avoid deleterious effects on the environment, the application has to be done in a limited area, and usually no more than twice per year. Moreover, the schedule of application is very tight, because of the shortage of helicopters available for this purpose. Above all, aerial spraying often misses the acute timing of beetle emergence.

Many attempts have been made to develop the use of biological agents for controlling immature *M. alternatus* in dead pine trees, including woodpeckers (Yui *et al.*, 1993), insect predators (Inoue, 1991; Enda, 1992; Ueda *et al.*, 1999), parasitic nematodes (Ogura & Kosaka, 1991) and fungal parasites (Shimazu, 1995). Though some of them are promising for practical use, there is no biological agent that has proved to successfully reduce the incidence of PWD in field conditions, so far.

For protecting individual pine trees from PWD infection, nematicide injection into tree bodies has been developed (Matsuura, 1980). The following nematicides are on the market and have come to be widely used in Japan; emamectin benzoate, levamisole HCl, mesulfenos, milbemectin, morantel tartrate, and nemadectin. The cost of nematicide application for a tree (30 cm DBH) amounts to 1000 yen on average in Japan.

Fukiage-hama seacoast pine forest and history of pine wilt damage

Fukiage-hama is one of the largest seacoast pine forests in Japan. It is located in southern part of Kyushu Island, southwestern Japan, and facing the East China Sea (Fig. 2). Fukiage-hama covers an area of 1550 ha and the length of the pine forest reaches as long as *ca* 30 km. The forest mainly consists of Japanese black pine, *Pinus thunbergii*, known to be highly susceptible to PWD, but widely planted on sand dunes because of high tolerance to wind and salt. The pine trees in Fukiage-hama were planted originally in the late 1800s after a devastating forest fire and have played an important role in protecting the local communities from strong winds from the East China Sea and the movement of sand.

In Fukiage-hama, damage by PWD has been reported since the 1940s. The damage, however, was not very severe prior to 1992 (Fig. 2), although conventional control methods including aerial spraying had been conducted after 1973. In 1991, there was a slight increase in the damage level caused by a powerful typhoon in September. In 1992, PWD damage increased sharply to over 10 000 m^3 (*ca* 800 m^3/ha). While conventional control operations had been conducted continuously, the damage level was not sufficiently suppressed. In 1994, the damage level jumped up to nearly 15 000 m^3 (*ca* 1100 m^3/ha). Over 50 000 pine trees were killed by PWD in 1994 alone.

The outbreak of PWD in Fukiage-hama after 1992 had a distinct pattern in the spread of the damaged area (Soné *et al.*, 1996) (Fig. 3). In 1992, dead trees were found mostly in the southern part, suggesting

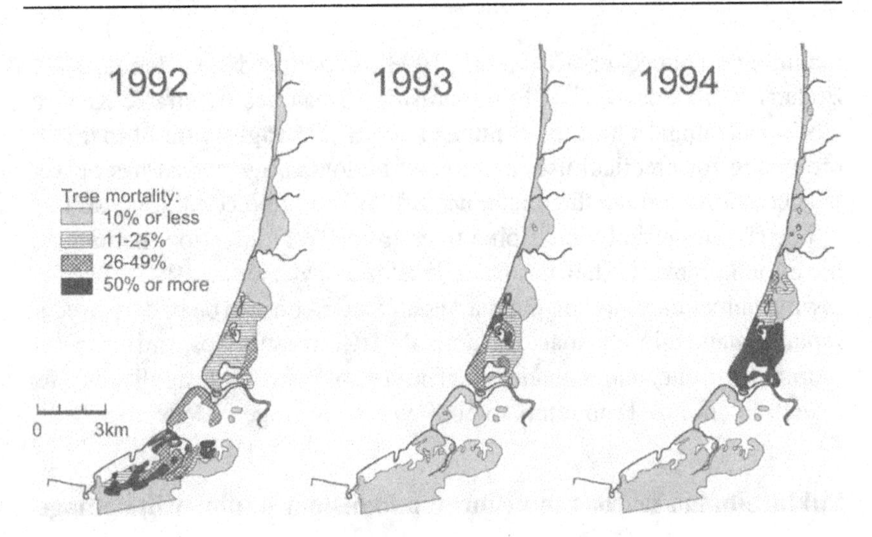

Fig. 3. *Spread of the damage area of pine wilt disease in Fukiage-hama (Soné* et al., *1996).*

that there was a source of infection around that area. In fact, we found severely damaged pine forest on the mountains in the eastern adjacent area of Fukiage-hama. In 1993, the epicentre moved to the north. In the southern part, the damage level decreased to less than 10%, meaning that no adult pine trees that could be infected remained. In that year, we had an abnormally cold summer with much precipitation. It seems that this climatic condition prevented the development of the disease. In contrast, we had an abnormally hot and dry summer in 1994: PWD raged again and the damaged area covered the middle part of Fukiage-hama. The area was also spreading to the north. In such a manner, the damaged area spread from the south to the north through Fukiage-hama, and the rate of disease spread was 2-5 km per year (Soné *et al.*, 1996).

WHY WAS THE SPREAD OF THE DISEASE NOT PREVENTED?

To cope with this outbreak, we needed to review the reason why PWD damage increased so sharply after 1992.

Meteorological events might have promoted the outbreak of PWD. The damage level in Fukiage-hama rose sharply in 1992, just after the strong typhoon struck in 1991. Tree stress caused by the typhoon damage seemed to be a background to the PWD outbreak in 1992. The high temperature and drought in the summer of 1994 also influenced the severity of PWD.

The outbreak began in the southern part of Fukiage-hama and was thought to have been brought by the insect vector invading from the adjacent damaged forest. It is likely that close observation of the source of infection and an effective control plan would have prevented an epidemic of the disease.

However, the most important factor of the failure in Fukiage-hama was the incomplete disposal of infested trees. In spring of 1993, we found huge amounts of infested logs waiting for transportation to the pulp mill and a considerable number of them were moved after insect emergence started. This was because the number of dead trees exceeded the upper limits of logging operations as well as that of the pulp mill. Branches unwanted by the pulp mill were mostly left in the forest and many of them had emergence holes of *M. alternatus*. Moreover, we often found emergence holes on the pesticide-treated trees that evidently showed the limited effect of the infiltrative pesticide. The fundamental problem with such technicalities was that there had not been enough manpower and funds to eradicate all infested trees in time.

A failure to synchronise the timing of aerial spraying and the emergence time of *M. alternatus* was thought to be the other important factor of the fiasco. It is very important to know the precise insect emergence time. Because adult emergence is highly susceptible to environmental conditions, it is preferable to carry out a census of adult emergence on site. In many cases in Japan, however, the census is done at a place convenient for the observer, *e.g.*, in a field cage at the research station. This was also true in Fukiage-hama until 1994. Moreover, the schedule of application for this area had conventionally been decided in January or February. This is because the fiscal year in Japan changes on April 1. The emergence time of *M. alternatus* is strongly influenced by the spring temperature. The early decision regarding application schedule prevented us from acutely adjusting application timing.

SPECIAL CONTROL OPERATIONS AFTER 1994

After the devastating damage of 1994, together with Kumamoto District Forest Office of the Forestry Agency, we drew up a plan to make control operations more effective. For improvement of the control operations, eradication of infested trees and adjustment of the timing of aerial spraying was promoted as a priority.

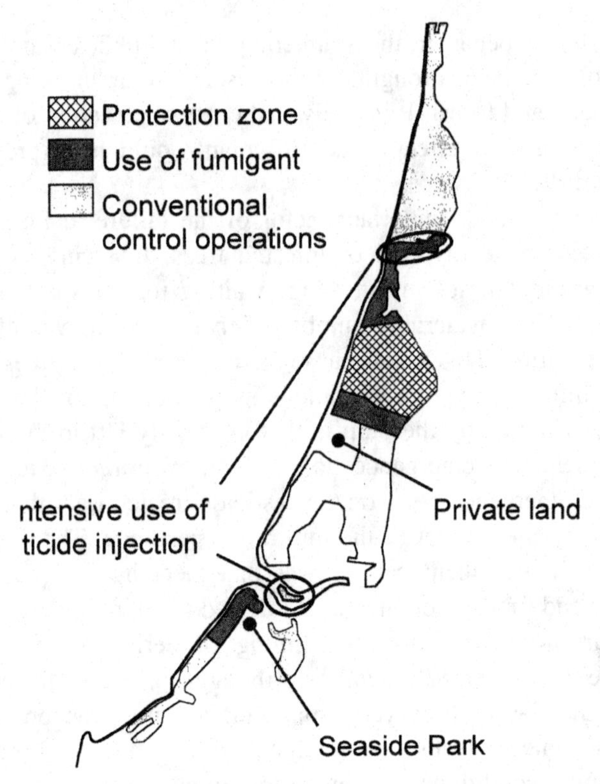

Fig. 4. *Zoning for effective assignment of control efforts to eradicate the infested trees in Fukiage-hama after 1994. A 'protection zone' was established adjacent to the epicentre in 1994, and complete eradication of the infested trees including branches was performed. Fumigant was used in and around the protection zone, and around the Seaside Park. Aerial spraying was also conducted over the whole area.*

Eradication of infested trees requires huge amounts of labour and money, and the manpower, as well as the budget, has its upper limits. Thus we planned to zone the pine forest to concentrate control efforts into the most critical area.

At that time, the damaged area was spreading from the middle to the northern part of Fukiage-hama. The pine forest in the southern part had almost been destroyed. Thus we created a 'protection zone' in the northern adjacent area to the epicentre in 1994 (Fig. 4). This zone was about 2-3 km wide, which corresponded to the yearly spread of the damaged area. In the protection zone, all dead trees including branches were eradicated completely. Fumigant was used extensively, instead of infiltrative pesticide. Fumigant was also used for the areas adjacent to the protection area and near the Seaside Park, but not for the whole area because of a limited budget. For speeding up disposal, machines such as fellers or tractors were introduced. A chipper machine to dispose of the branches was also introduced. Other machines such as power shovels or bulldozers were used to bury the logs underground. Forest roads were newly constructed for carrying out logs for the pulp mill and for movement of the large machines. As a result, we managed to finish eradicating the infested trees before the insect emergence season.

To estimate adult emergence time acutely, we conducted a census of *M. alternatus* emergence on site. Data from the census served for making plans for aerial spraying and estimating the effect of the application. Through the close coordination with administrative sections, the decision-making of the application schedule was put back to April, and the additional application of insecticide (three times per year) was carried out in 1995 and 1996. With those adjustments, we could cover the beetle emergence time with the residual time of the sprayed insecticide.

Nematicide injection was used for remarkable pine trees around the protection zone and in a part of the Seaside Park to the limit of the budget. A small number of surviving trees in the damaged area that can become the seed source of the regenerated pine stand were also treated with the nematicide.

We also tried to decrease the potential source of infection in Fukiage-hama. In the devastated area, planting of the resistant clone of *P. thunbergii* or broad-leaved trees was promoted. Those trees will not be subjected to PWD, and therefore not become new sources of infection.

Results of the special control operations

As a result of the special control operations from 1994 to 1996, the outbreak of PWD almost subsided by 1997 (Fig. 1). The spread of the damaged area (Fig. 3) towards the north was checked and a large area of pine forest in the northern part of Fukiage-hama has been saved from the epidemic. The amount of newly diseased trees in 1995 was, however, still about half that in 1994 despite the complete eradication of infested logs, suggesting the latent infection of *B. xylophilus* in healthy-looking trees (Futai, 1995). This fact tells us that a single year's effort against PWD outbreak should prove fruitless.

Table 1. *Costs of control operation against pine wilt disease in Fukiage-hama.*

Item of expense	Yearly expenditure (× 1000 JPY)						
	1993	1994	1995	1996	1997	1998	1999
Disposal of infested trees	39 250	138 380	220 378	64 440	16 590	4590	0
Aerial spraying	65 110	67 660	145 380	131 104	63 386	62 404	62 332
Nematicide injection	980	8490	41 240	29 004	0	3884	3884
Other operations	19 800	69 540	196 180	94 788	114 376	159 492	136 822
Total	125 140	284 070	603 178	319 336	194 352	230 370	203 138

It cost huge amounts of money to perform the special control operations (Table 1). The amount of money used in 1994 was 284 million yen, over twice that in 1993. A total of 603 million yen (equivalent to *ca* 5 million US$) was spent for a single year in 1995. The amount spent decreased as the damage level dropped, but still *ca* 200 million yen was spent in 1999 mainly for reforestation of the devastated area.

Although it was intended to decrease the potential source of infection by planting PWD-resistant pine seedlings or broad-leaved trees, an extensive area of the destroyed pine stand has been covered by naturally regenerated *P. thunbergii* seedlings. Those trees should be susceptible to PWD. Additionally, there are privately owned pine forests within and around Fukiage-hama, in which the mandatory eradication of the infested trees is not always applicable. Consequently, elimination of the potential source of infection around Fukiage-hama has not been completed. Careful precautions against the next outbreak of PWD are still needed in Fukiage-hama.

Key to successful control of pine wilt disease

Here we summarise control operations along the stage of the epidemiology, with lessons from the PWD outbreak in Fukiage-hama.

In the areas where PWD has not yet invaded, top priority should be given to quarantine. It may be costly, but is expected to be far less expensive than control operations after invasion, as shown in Table 1. Stress factors may promote the occurrence and development of PWD, so removing such factors could be helpful in preventing the occurrence of PWD to some extent. Many of those factors, however, are out of our control such as typhoons or drought in summer. Once the disease has spread, removal of stress factors itself seems not effective enough because of the strong virulence of the pathogen.

It is probable that complete eradication of PWD is feasible only at the very first stage after invasion. At this stage, however, it is very difficult to find the diseased pine trees to eradicate because of their small number. Aerial searching for dead trees could be helpful to overcome this difficulty. If the small number of initial dead trees is overlooked, PWD will be prevalent soon with its strong virulence.

In an outbreak stage such as Fukiage-hama before 1995, zoning for effective concentration of control efforts should be useful, especially under conditions of limited budget and manpower. In the area where the control effort will be made a priority, complete eradication of infested trees before insect emergence time must be performed. If conditions allow, it is worth considering introducing machinery to improve disposal. In such a situation, a forest road is an important infrastructure to run the control operation. We should be prepared to spend at least 2 consecutive years of control efforts to suppress the outbreak, because there could be trees latent infected by PWD.

Aerial spraying is a powerful tool for controlling PWD, but we should be most careful to adjust the timing of application to hit the emergence time of the insect vector. Thus, this method requires precise understanding of the ecology of the insect. If the insect emergence time is not limited to a particular part of the year, the application of this method would not be practical. Also it is important to recognise that this method is designed for preventing the disease and never exterminates the insect vectors by itself.

After suppression of the outbreak, watching for new sources of infection will be imperative. Isolation of the pine stand from the possible source will minimise the risk of reinfection and make the control operation easy in case of a future outbreak. This could be carried out by destroying the adjacent unmanaged pine stands, or converting them into PWD-resistant clones of pine trees or broad-leaved trees.

Acknowledgement

We thank Mr H. Shimizu and Mr K. Yamabe of Kyushu District Forest Office of the Forestry Agency, Japan, for their arrangements for the field work and for providing the administrative statistics.

References

ENDA, N. (1992). [Control of *Monochamus alternatus* using parasitic wasp *Scleroderma guani* in China.] *Shinrin-boeki [Forest Pests]* 41, 126-131.

FUTAI, K. (1995). The role of the symptomless carrier in epidemic spread of the pine wilt disease. *International symposium on pine wilt disease caused by pine wood nematode*, pp. 69-80.

INOUE, E. (1991). [Studies on the natural enemy of *Monochamus alternatus* Hope, *Dastarcus longulus* Sharp (Coleoptera: Colydiidae).] *Bulletin of Okayama Prefectural Forestry Experiment Station* 10, 40-47.

KISHI, Y. (1995). *The pine wood nematode and the Japanese pine sawyer.* Tokyo, Japan, Thomas Company, 302 pp.

KIYOHARA, T. & TOKUSHIGE, Y. (1971). [Inoculation experiments of a nematode, *Bursaphelenchus* sp. onto pine trees.] *Journal of the Japanese Forestry Society* 53, 210-218.

KOBAYASHI, F., ENDA, N. & TABATA, K. (1981). [Effect of insecticide sprayed in winter and spring on *Monochamus alternatus* in dead pine logs.] *Transactions of the 92nd Annual Meeting of the Japanese Forestry Society*, pp. 369-370.

KOBAYASHI, F., YAMANE, A. & IKEDA, T. (1984). The Japanese pine sawyer beetle as the vector of pine wilt disease. *Annual Review of Entomology* 29, 115-135.

MAMIYA, Y. (1988). History of pine wilt disease in Japan. *Journal of Nematology* 20, 219-226.

MAMIYA, Y. & ENDA, N. (1972). Transmission of *Bursaphelenchus lignicolus* (Nematoda: Aphelenchoididae) by *Monochamus alternatus* (Coleoptera: Cerambycidae). *Nematologica* 18, 159-162.

MATSUURA, K. (1980). [Control of the pine wilt disease caused by pine-wood nematodes with trunk-injection.] *Shokubutsu-boeki [Plant Protection]* 38, 27-31.

MORIMOTO, K. & IWASAKI, A. (1972). [Role of *Monochamus alternatus* (Coleoptera: Cerambycidae) as a vector of *Bursaphelenchus lignicolus* (Nematoda: Aphelenchoididae).] *Journal of the Japanese Forestry Society* 53, 210-218.

NAKANE, I. (1977). [Tests for the effectiveness of insecticide applied to dead pine logs infested by *Monochamus alternatus*.] *Bulletin of Hiroshima Prefectural Forestry Experiment Station* 12, 49-83.

OGRA, N. & KOSAKA, H. (1991). Biology of a tylenchid nematode parasitic on the Japanese pine sawyer, *Monochamus alternatus*. *Nematologica* 37, 455-469.

SHIMAZU, M., TSUCHIYA, D., SATO, H. & KUSHIDA, T. (1995). Microbial control of *Monochamus alternatus* Hope (Coleoptera: Cerambycidae) by application of nonwoven fabric strips with *Beauveria bassiana* (Deutyeromycotina: Hyphomycetes) on infested tree trunks. *Applied Entomology and Zoology* 30, 207-213.

SONÉ, K., IZUMI, A. & HAYASHI, S. (1996). [Mass mortality of pine trees at the Fukiagehama National Forest.] *Bulletin of the Faculty of Agriculture Kagoshima University* 46, 1-8.

TSUCHIYA, T. (1985). [Eradication of *Monochamus alternatus* by fumigant.] *Shinrin-boeki [Forest Pests]* 34, 123-127.

UEDA, A., FUJITA, K., URANO, T. & YAMADA, M. (1999). [Effect of released adults of *Trogassita japonica* (Coleoptera: Trogossitidae) on the Japanese pine sawyer, *Monochamus alternatus* (Coleoptera: Cerambycidae) in pine bolts.] *Applied Forest Science* 8, 169-172.

YUI, M., SUZUKI, S. & NAKAMURA, M. (1993). [Woodpeckers as a predator of *Monochamus alternatus* and their preservation.] *Shinrin-boeki [Forest Pests]* 42, 105-109.

Nematology Monographs & Perspectives, 2003, Vol. 1, 283-289

Pathogens of the pine sawyer, *Monochamus alternatus*, in China

Lai Fa WANG [1], Fu Yuan XU [2], Li Ya JIANG [3], Pei ZHANG [2] and Zhong Qi YANG [1]

[1] *The Chinese Academy of Forestry, Beijing, 100091 China*
[2] *Forestry Academy of Jiangsu, Nanjing, Jiangsu, 211153 China*
[3] *General Center for Forest Pest Control of Anhui Province, Hefei, Anhui, 230031 China*

Summary – A survey of pathogens of *Monochamus alternatus*, a vector of *Bursaphelenchus xylophilus*, was carried out in China. Ten entomogenous fungal species and one bacterial species were identified on *M. alternatus* in pine forests in Anhui and Jiangsu provinces, of which *Aspergillus flavus*, *Alternaria alternata*, *Fusarium moniliforme*, *F. oxysporum*, *F. solani*, *Paecilomyces farinosus*, *Verticillium lecanii* and *Serratia marcescens* were new records on *M. alternatus* in China and *Beauvernia bassiana* was the prominent species. *Fusarium oxysporum*, *A. flavus*, *B. brongniartii* and *S. marcescens* were frequent species. Two different species of fungi or one species of fungus and one species of bacterium were found on the same *M. alternatus* larva. About 1-5% of larvae were diseased or infected by the pathogens.

It is well known that the pine wilt disease, caused by the pine wood nematode (PWN), *Bursaphelenchus xylophilus* (Steiner & Buhrer) Nickle, has been one of the most serious problems for pine (*Pinus* spp.) forests in China. At present, no economic and practical control method has been found at home or abroad. As the pine sawyer, *Monochamus alternatus* Hope is the vector of PWN and is found in 20 provinces in China, it is vitally important to control the pine sawyer to prevent this disease. Reducing the population density of *M. alternatus* can be put into effect with comprehensive measures, such as removing dead and weak trees, and using chemical and biological methods. Meanwhile, in natural populations of *M. alternatus*, death caused by contagious diseases was observed at every stage and is supposed to be an important component of the total population mortality. Application of pathogens to control *M. alternatus* may be a desirable approach to practical nematode management. In China, some investigative work on fungi of the pine sawyer in Nanjing city, Jiangsu province, was done by Zhou (1995), with the entomogenous fungi *Beauveria bassiana* (Balsama) Vuillemin, *B. brongniartii* (Scardo) Petcl., *Metarrhizium anisopliae* (Metsch) Sorokin and *Acremonium* sp. Survey of pathogens of the pine sawyer is important and basic for biological control, hence this survey was made in PWN areas of Jiangsu and Anhui provinces in March and June, 2001. This is a preliminary report.

Materials and methods

Origins of samples

The samples were inactive larvae and cadavers of *M. alternatus* collected from various pine forests infected by PWN. Each inactive larva was kept at 25°C in a test tube sealed with an absorbent cotton pellet until fungal filaments appeared on the bodies of the larvae.

Isolation of microorganisms

Potato dextrose agar (PDA) plates or nutrient agar plates containing 1% yeast extract were used for isolating fungi and bacteria, respectively. All the isolates were cultured and purified and stored in a refrigerator.

Identification of isolates

Identification of fungal isolates was verified by reference to Domsch and Gams (1980), Nelson *et al.* (1983) and Pu and Li (1992). Bacterial identification was verified by reference to Buchanan and Gibbons (1974).

Results

Identification and descriptions of isolates

Ten entomogenous fungal species and one bacterial species on *M. alternatus* in pine forests in Anhui and Jiangsu provinces were obtained and identified. The species are as follows.

Alternaria alternata *(Fr.) Keissler*

Colonies growing fast, brown or black, reaching 6.5 cm diam. in 7 days at 25°C on PDA. Conidiophores usually straight or curved, one to three septate, to 50 μm long, 3-6 μm wide. Conidia ovoid, obclavate obpyriform, 18-60 × 7-18 μm, three to eight transverse septa, with or without a short conical or cylindrical apical beak not exceeding one third of the conidial length.

Aspergillus flavus *Link ex Gray*

Colonies growing fast, yellow, reaching 5.0-6.4 cm diam. in 7 days at 25°C on PDA. Conidiophores 0.4-1.0 cm long and rough-walled. Conidial heads radiating; on larger condiophores a layer of metulae supports the philalides, whilst on smaller ones metulae are absent. Conidia globose to subglobose, 3.5-4.5 μm diam.

Beauveria bassiana *(Balsamo) Vuillemin*

Colonies growing slowly, reaching 0.8-1.0 cm diam. in 7 days at 25°C on PDA, woolly, often appearing powdery, at first white but later some becoming yellow. Conidiogenous apparatus forming dense clusters of swollen stalk cells which consist of a subglobose to flask-shaped venter, 3-6 × 2.5-3.5 μm, and a zig-zag shaped, denticulate rhachis to 25 μm long and 1 μm wide. Conidia hyaline, smooth-walled, mostly globose, 2-3 μm diam.

Beauveria brongniartii *(Sacardo) Petcl*

Colonies as in *B. bassiana*, reaching 0.8-0.9 cm diam. in 7 days at 25°C on PDA. Conidiogenous cells with a flask-shaped to subcylindrical venter, 4-15 × 2.0-3.5 μm, and a zig-zag shaped denticulate rhachis to 25 μm long and 1.0-1.5 μm wide. Conidia hyaline, smooth-walled, ellipsoidal, 2.5-4.5 × 2.0-2.5 μm.

Fusarium moniliforme *Sheldon*

Colonies reaching 6.0-6.5 cm diam. in 7 days at 25°C on PDA. Aerial mycelium sparse to abundant, white with purple. Conidiophores unbranched and branched monophialides. Micro-conidia abundant, head or in chains, 0-septate, oval, 7.1-11.5 × 1.5-2.5 μm. Macro-conidia generally abundant, 3-5 septate, slightly sickle-shaped, 24.3-50.0 × 2.7.0-3.8 μm. Chlamydospores absent.

Fusarium oxysporum *Schlechet. emend. Snyd. & Hans.*

Colonies growing fast, reaching 7.0-7.5 cm diam. in 7 days at 25°C on PDA. Aerial mycelium sparse to abundant, white with purple or blue. Conidiophores unbranched and branched monophialides. Micro-conidia generally abundant, mostly 0-septate, oval, ellipsoidal and cylindrical, straight or curved, 5.4-12.0 × 2.3-3.5 μm. Macro-conidia generally abundant, 3-5 septate, slightly sickle-shaped, 27.0-45.0 × 3.0-4.5 μm. Chlamydospores single or in pairs.

Fusarium solani *(Mart.) Sacc.*

Colonies reaching 5.5-6.0 cm diam. in 7 days at 25°C on PDA. Conidiophores unbranched and branched monophialides. Micro-conidia abundant, in chains, 0-septate, oval to club-shaped with a flattened base. 5.4-12.0 × 2.3-3.5 μm. Macro-conidia abundant, stout, thick-walled, and generally cylindrical, mostly 3-septate, 28.0-42.0 × 4.0-4.0 μm. Chlamydospores single or in pairs. The monophialides bearing micro-conidia are long when compared to those in *F. oxysporum.*

Paecilomyces farinosus *(Holm. Ex Gray) Brown & Smith*

Colonies growing moderately fast, reaching 1.9-2.8 cm diam. in 7 days at 25°C on PDA, white or bright yellow, sometimes with conspicuous yellow synnemata. Conidiophores usually erect, 100-300 μm tall, bearing several whorls of flask-shaped phialides. Conidia ellipsoidal to fusiform, 2.0-3.0 × 1.0-1.8 μm.

Metarrhizium anisopliae *(Metsch.) Sorokin*

Colonies reaching 2.0 cm diam. in 7 days at 25°C on PDA, white or bright yellow, sometimes with conspicuous yellow synnemata. Phialoconidia compacted into regular chains and columns. The column formation is due only to the aggregation of the elongate conidia themselves. Conidia 4.5-7.5 × 2.5-3.1 μm.

Verticillium lecanii *(Zimmermann) Viegas*

Colonies reaching 2.5-2.8 cm diam. in 14 days at 25°C on PDA, white to pale yellow, rarely with fasciculate hyphae, reverse uncoloured or yellow. Phialides solitary or in scant whorls arising from erect conidiophores or little differentiated prostrate aerial hyphae, aculate, 15.0-39.0 × 0.8-2.5 μm. Conidia in heads or parallel bundles, cylindrical with rounded tips or ellipsoidal, 2.3-10.0 × 1.0-2.6 μm. Chlamydospores absent.

Serratia marcescens *Bizio*

Cell straight rod, 0.5-1.0 × 0.4-0.5 μm, rod around with many flagella. Gram staining negative; colonies producing pink pigment.

COEXISTENCE OF FUNGI AND BACTERIA ON THE SAME BODIES OF *M. ALTERNATUS*

In March, the isolation results of 141 samples from Jiangsu showed that there were 22 bacteria isolates (13 *Serratia marcescens* and nine other bacteria) coexisting with fungal isolates in the same *M. alternatus* bodies. These fungi were *A. flavus*, *P. farinosus*, *B. bassiana* and *Fusarium* spp. Two fungal isolates also appeared in 19 samples, in which the combinations of *A. flavus* and *Fusarium* spp., *A. flavus* and *B. bassiana*, *Fusarium* spp. and *B. bassiana* coexisted on the same bodies.

FREQUENCIES OF FUNGI AND BACTERIA

Ten fungal species and one bacterial species were identified from 309 isolates from *M. alternatus* from the PWN areas of Jiangsu and Anhui provinces, from which *Aspergillus flavus*, *Alternaria alternata*, *Fusarium moniliforme*, *F. oxysporum*, *F. solani*, *Paecilomyces farinosus*, *Verticillium lecanii* and *Serratia marcescens* were new records on *M. alternatus* in China.

From Table 1, *B. bassiana* was the prominent species, and *F. oxysporum*, *Aspergillius flavus*, *B. brongniartii* and *S. marcescens* were frequent spccics.

The numbers and frequencies of entomogenous fungi may differ due to time and site differences, as in samples taken in March and June, from the same site, Tie Xin Qiao village of Nanjing (Table 2).

Comparing the numbers and frequency of entomogenous fungi on *M. alternatus* in Jiangsu with that in Anhui, *Aspergillus flavus* and *Alternaria alternata* had not been found in Anhui, and the frequency of *Beauveria bassiana* in Anhui (25%) was more than that in Jiangsu (17.5%).

In nature, about 1-5% of *M. alternatus* larvae were diseased or infected by the pathogens. At some sites, the infection rate was higher (17%).

Table 1. *List of fungi and bacteria isolated from* Monochamus alternatus.

Fungal species	Number of isolations	Frequency (%)
Beauveria bassiana	59	19.1
B. brongniartii	19	6.1
Aspergillus flavus	49	15.9
Alternaria alternata	2	0.64
Fusarium moniliforme	9	2.6
F. oxysporum	46	14.9
F. solani	14	4.5
F. spp.	33	10.7
Paecilomyces farinosus	7	2.3
Metarrhizium anisopliae	4	1.2
Verticillium lecanii	8	2.3
Other fungi	19	6.2
Serratia marcescens	29	9.4
Other bacteria	13	4.2
Total	309	100

Table 2. *Fungi species and frequencies on* Monochamus alternatus *sampled on two occasions from Tie Xin Qiao village, Nanjing.*

Fungal species	March		June	
	Numbers	Frequency	Numbers	Frequency
Beauveria bassiana	8	13.2	7	12.1
B. brongniartii	1	1.6	2	3.4
Aspergillus flavus	27	44.2	7	12.1
Alternaria alternata	0	0	2	3.4
Fusarium moniliforme	3	4.9	2	3.4
F. oxysporum	6	9.9	11	19.1
F. solani	2	3.3	0	0
F. spp.	11	18	17	29.3
Paecilomyces farinosus	0	0	2	3.4
Metarrhizium anisopliae	1	1.6	3	5.2
Verticillium lecanii	2	3.3	5	8.6
Total	61	100	58	100

Discussion

Natural enemy species can vary greatly in attributes that affect their ability to control a particular target, even among different populations of the same species. This suggests that selection or breeding of isolates is necessary. The pathogenicity of most isolates to *M. alternatus* examined by the authors needs to be tested later. However, there is little knowledge of some aspects, such as the epidemiology and ecology of entomogenous pathogens infecting *M. alternatus* and mode of action of weak pathogenetic *Fusarium* spp. and *S. marcescens* and the relation of pathogenetic fungi and bacteria, so this survey of pathogens to *M. alternatus* in nature may help develop biocontrol.

References

BUCHANAN, R.E. & GIBBONS, N.E. (EDS) (1974). *Bergey's manual of determinative bacteriology.* Baltimore, USA, The Williams & Wilkins Co., 1246 pp.

DOMSCH, K.H., GAMS, W. & ANDERSON, T.H. (1980). *Compendium of soil fungi.* Two volumes. New York, NY, USA, Academic Press, 809 pp.

NELSON, P.E., TOUSSON, T.A. & MARASSAS, W.F.O. (1983). Fusarium *species: an illustrated manual for identification.* Pennsylvania, USA, Penn. State University Press, 193 pp.

PU, Z.L. & LI, Z.Z. (1992). *Insect Mycology.* Hefei, Anhui, China, Anhui Publishing House of Science and Technology, 715 pp.

XING-HENG, Z. (1995). Investigation on entomogenous fungi of the pine sawyer in Nanjing city. In: Yang, B.J. *et al.* (Eds). *Epidemiology and Management of the Pinewood nematode Disease in China*, Chinese Publishing House of Forestry, pp. 188-191.

Nematology Monographs & Perspectives, 2003, Vol. 1, 291

Activity of avermectin for controlling the pine wood nematode, *Bursaphelenchus xylophilus*

Mao Song LIN and Ming-guo ZHOU

Department of Plant Protection, Nanjing Agricultural University, Nanjing, 210095 China

Summary – When avermectin antibiotic was applied by trunk injection at 60-90 ml/plant to *Pinus thunbergii* and *P. massoniana*, a residual efficacy was achieved for at least 2 years for control of *B. xylophilus.* After treatment, residual control by avermectin injection was 100% in the 1st year and 88% in the 2nd year. Avermectin injected into *P. thunbergii* spread to the top of the tree, as shown by inspecting nematodes from treated wood. When a large quantity (210 ml/plant) of 2% avermectin was used, no phytotoxicity was observed to affect *P. thunbergii* growth. Application by trunk injection was a safe alternative method for control of nematodes with reduced risks to the environment and to people.

Printed in the United States
by Baker & Taylor Publisher Services